Wilson Popenoe

Agricultural Explorer, Educator, and Friend of Latin America

by Frederic Rosengarten, Jr.

A PUBLICATION OF
NATIONAL TROPICAL BOTANICAL GARDEN
LAWAI, KAUAI, HAWAII 96765

TO M.O.R.

Contents

ISBN 0-935868-53-4
Library of Congress Catalog Card Number 91-062018

Printed in the United States of America

Preface

I met Wilson Popenoe for the first time in March 1940 in Guatemala. He was forty-eight; I was twenty-three. He had just been "loaned" by the United Fruit Company to Merck and Co., Inc., for six months to take charge of the Cinchona (quinine) project of Merck's subsidiary in Guatemala. I was to be his assistant. This was my first job, and he was my first boss—a unique educational opportunity for a young man.

During 1940, I traveled extensively throughout Guatemala with Dr. Popenoe as we encouraged Guatemalan planters to grow cinchona trees in order to increase the production of quinine in the Western Hemisphere. He taught me a lot about Central America, tropical agriculture, economic botany, and many other things. A witty raconteur, he used to tell amusing jokes in English and Spanish while we bumped along the rough roads for hours on end.

Now, fifty years later, it is my privilege to write a brief life history of Wilson Popenoe—the agricultural explorer, educator, author, horticulturist, and humanitarian. Apart from his scientific and educational achievements, he was a most skillful and "simpático" ambassador of good will for the United States in Central America. He knew how to get along with everybody in the region. The humble plantation laborer, the wealthy landowner, the student, even the most intractable government official soon became his friend.

Acknowledgments

The author acknowledges his sincere appreciation for the help given him by many persons during the preparation of this book. Wholehearted thanks are due to Dr. Hugh Popenoe, son of Wilson Popenoe, Director of the Center for Tropical Agriculture at the University of Florida, who provided much encouragement, information, constructive suggestions, and many photographs; to Nancy Popenoe Guillou of Atherton, California, and Dr. Marion Popenoe Hatch, of Antigua, Guatemala, daughters of Wilson Popenoe; and to his nephews Dr. John Popenoe, former director of the Fairchild Tropical Garden, Miami; Dr. David Popenoe of Princeton, New Jersey; and Paul Popenoe of Lafayette, California—for providing photographs and records of the Popenoe family and Wilson Popenoe. Thanks are also due to the estate of Wilson Popenoe for granting permission to reproduce the photographs on pages 9, 13, 27, 50, 52, and 108 of this book, which appeared originally in *Manual of Tropical and Subtropical Fruits* by Wilson Popenoe, published in 1920 by the Macmillan Company, New York.

The author is most grateful to Dr. Simón E. Malo, director of Escuela Agrícola Panamericana, Zamorano, Honduras, for thoughtful criticism and information concerning the chapter about that school; to Mrs. Anita Karg, archivist, and the staff at the Hunt Institute for Botanical Documentation, Carnegie Mellon University, Pittsburgh, for providing numerous photographs and other useful research data, including unpublished autobiographical notes of Wilson Popenoe; to Kate Semerad and Mimi Britt of the Washington, D.C., office of Zamorano; and to Ned Rimer of Zamorano, Honduras, for many school photographs, annual reports, and other useful data.

The author acknowledges with thanks the cooperation of the following individuals who provided him with photographs and information or otherwise assisted him during the course of the work: Robert Armour of Reston, Virginia; Dr. Robert P. Bates and Dr. Nigel J. H. Smith of the University of Florida; Jorge Benítez, of Antigua, Guatemala; Jean Boise, of the Botany Libraries, Arnold Arboretum, Harvard University; Alfred L. Bush of the Princeton University Library; Carolyn F. Clewell and Barbara Shattuck of the National Geographic

Society, Washington, D.C.; Dr. Raymond Crist of Gainesville, Florida; Dr. Joseph Ewan of the Missouri Botanical Garden; Lizz Frost, Baker Library, Harvard Graduate School of Business Administration; George P. Gardner of Boston; Mrs. Jayne MacLean, National Agricultural Library, Beltsville, Maryland; Mary Francis Morrow, National Archives, Washington, D.C.; Amado Pelén of Zamorano, Honduras; John H. R. Plews of Honolulu; Jack Shepherd, California Avocado Society, Altadena, California; the late Nixon Smiley of Miami, Florida; John G. Smith of Guatemala City; Clyde S. Stephens and John D. Dickson of Tela, Honduras; Dr. William L. Theobald and Dr. David H. Lorence, National Tropical Botanical Garden, Lawai, Kauai, Hawaii; Dr. Thomas L. Welch, Columbus Memorial Library, Washington, D.C.; Bertram Zuckerman of the Fairchild Tropical Garden, Miami; and the staff of the American Heritage Center, University of Wyoming.

The author is thankful to Barbara Westergaard for her skillful editing.

Wilson Popenoe. 1892–1975. (*Courtesy of California Avocado Society.*)

Chapter I

Early Years, 1892–1913

The first Popenoe to arrive in the New World was Jean Papineau who came to Boston about 1690 with a group of Huguenots, who may have come from the region of La Rochelle, a seaport in western France on the Bay of Biscay. By 1697 Jean was living in New Oxford, Massachusetts, a vulnerable frontier colony to which alien settlers were sent by the Massachusetts authorities to serve as a protective buffer against the attacks of hostile Indians. He was in the "wash leather" business, purchasing raw skins from the Indians and settlers, dressing them, then shipping the cleansed skins by wagon to be sold to hatters in Newport and Boston.

Following the Deerfield massacre of 1704, when a band of Indians burned that town and killed many colonists, New Oxford was abandoned. Most of the settlers, including Jean Papineau, moved to New Rochelle, a Huguenot colony near New York City. Soon thereafter Jean married a French girl named Charlotte Bounos.

In 1706 their son Pierre Papineau was born. Some years later, when they moved to New Jersey, the spelling was changed to Popino. The name may be translated as "scanty beard." It is derived from the medieval French *pappins* and the Latin *papus,* meaning down or pubescence.

A son, Peter Popino, emigrated to the western part of Virginia near Morgantown and fought with the American Revolutionary army against the British in 1777, then moved to a farm in Kentucky. Following the Revolution, the government of Virginia gave large parcels of land in Kentucky to its Revolutionary soldiers. According to family tradition, Peter was killed in 1790 near Boonesboro, on the Kentucky River, the site of a fort founded by Daniel Boone in 1775. Peter was apparently the victim of a surprise attack by the Indians—shot while plowing his field. It was said this Indian massacre was so ferocious every woman as well as every man had to shoulder a musket.

About this time, the spelling of the name of this branch of the family was changed again, to Popenoe. James Popenoe, the eldest son

Wilson Popenoe in Topeka, Kansas, 1896, age four. (*Courtesy of Hunt Institute for Botanical Documentation, Carnegie Mellon University, Pittsburgh, Pa.*)

of the above-mentioned Peter, was born in 1777 at Morgantown. He became a farmer, a soldier in the war of 1812, a sheriff, a tax collector, and a Jacksonian Democratic politician. In 1793 he moved with his family to Xenia, Ohio, and married Jane Davis, daughter of a Revolutionary veteran. After her death James married a second wife, Sarah Holcomb, who bore him eight children including Willis Parkinson Popenoe, born in 1826. Willis married Miranda Holcomb at Springfield, Illinois, in 1851. They moved to Topeka, Kansas, where they were among the first settlers. Willis was a successful farmer and merchant in Topeka. His fifth child, Frederick Oliver Popenoe, was born in 1863. Frederick, known as Fred, or "F.O.P." by his children, married Marion Amanda Bowman from Boston in 1887. Their children were: Paul Bowman Popenoe, born in 1888; Frederick Wilson Popenoe (known as Wilson and the subject of this biography), born March 9, 1892; and Herbert Franklin Popenoe, born in 1899.

As a young man, Fred learned stenography in Topeka, worked with the Santa Fe Railway, became a court reporter, and then worked as

Marion Bowman Popenoe, Wilson's mother, about 1890. (*Courtesy of Hunt Institute for Botanical Documentation, Carnegie Mellon University, Pittsburgh, Pa.*)

a private clerk for the governor of Kansas, John P. St. John. Fred subsequently went into real estate, banking, and the newspaper business. His wife, Marion, provided him with a substantial dowry; she was one of the New England Bowmans who were prosperous textile producers and devout Protestant Congregationalists.

Fred Popenoe became president of the Accounting Trust Company of Topeka, and for a time was owner and publisher of the newspaper

Wilson's father, Fred Popenoe, at the mining company office, San José, Costa Rica, 1902. (*Courtesy of Nancy Popenoe Guillou.*)

Topeka Daily Capital. Active in temperance work, he was also secretary of the National Committee of Anti-Saloon Republicans. In 1895, Fred built one of the finest homes in Topeka; he seemed to be a highly successful entrepreneur.

Despite his business accomplishments in Kansas, Fred became restless. When a friend in Topeka called his attention to a lucrative gold mine in Costa Rica that one could invest in, he decided to go down

Topeka State Journal, June 21, 1899. There was a lot of optimistic publicity in Kansas about the Bella Vista gold mine in Costa Rica before it failed. Wilson's father was one of the principal investors. (*Courtesy of American Heritage Center, University of Wyoming.*)

to Central America and look it over. Favorably impressed, Fred made a substantial investment in the Bella Vista Mine, located in the province of Puntarenas in west-central Costa Rica. Between 1896 and 1907 he made several long trips to Costa Rica in connection with gold mines. In 1901 he brought his family to Costa Rica to live with him for a year.

On April 4, 1897, upon his return from Costa Rica to Kansas, the *Topeka Daily Capital* interviewed Fred Popenoe, who made the following comments:

> The Costa Rican people are friendly, honest, peaceable, and never do today what they can possibly put off until tomorrow. The mineral wealth of Costa Rica is extensive, and, so far, untouched. A fortune can be made in that country by anyone who has some capital to begin with, plenty of judgment and perseverance to continue with and a willingness to put up with untold inconveniences: no others need apply.

At first the Costa Rican investment was called the Thayer Mining and Milling Company. In 1901 Fred changed the company name to

At the Bella Vista mine in Costa Rica, 1899. Fred Popenoe is second from right, front row. (*Courtesy of John Popenoe.*)

Río Grande Gold Mining Company and became its president. Río Grande owned and operated another gold mine called "El Porvenir" on the Río Machuca in the Aguacate Mountain range in Costa Rica.

By 1906 the gold mining business in Costa Rica was doing badly. Although Fred Popenoe had already invested a lot of his own and his wife's money in Costa Rica, he was obliged to borrow fifteen thousand dollars more from banks in Kansas just to keep the mines working. Unfortunately, things went from bad to worse. His health went downhill along with his finances. By 1907, he decided to cut his losses, sell out, and leave Costa Rica once and for all. He had already moved his family from Kansas to California in 1904, so he joined them there. Other businessmen in Topeka also lost money as a result of the losses incurred by Fred Popenoe in Costa Rica. That may be one of the reasons he decided never to return to Topeka.

Many years later Wilson observed it was lucky for him that his father's mining venture had failed in Costa Rica. Otherwise, he said, he, Wilson, would probably have been a lazy, wealthy bum, and never have done a stroke of work the rest of his life.

Wilson Popenoe was nine years old when the family lived in Costa Rica. He fell in love with Central America: with the hospitable people, the mild climate, the Spanish language, the exuberant forms of tropical plant life, the whole charming Latin modus vivendi. It was in Costa

Wilson in Costa Rica, 1901, age nine. (*Courtesy of Nancy Popenoe Guillou.*)

Rica in 1901 that young Wilson first came into contact with the avocado, or alligator pear, a fruit that was to play an important role for him later. He had inherited an interest in horticulture. His grandfather Willis P. Popenoe had been an avid fruit grower in Kansas, who in 1876 sent some of his own selected apples to be exhibited at the Philadelphia Centennial Exposition.

In the summer of 1904 Fred Popenoe had moved his wife and children from Topeka to Los Angeles; then in 1906 to Altadena, near

Charles Fuller Baker (1872–1927). Baker, a professor of botany at Pomona College in 1910, encouraged Wilson Popenoe to pursue a career in tropical agriculture. (*Courtesy of Hunt Institute for Botanical Documentation, Carnegie Mellon University, Pittsburgh, Pa.*)

Pasadena. During this period Wilson spent part of his time at Pasadena High School and the rest working in gardening and horticulture. He said himself he spent more time propagating plants than burning the midnight oil studying. He got excellent marks in the natural sciences, but did not do well in other subjects.

Following his graduation from Pasadena High School in 1910, Wilson learned that Charles Fuller Baker, a noted entomologist and botanist, was teaching at nearby Pomona College. Experienced in Latin America, Baker had taken part in an expedition to Colombia to collect tropical plants and insects, served as chief botanist at an experiment station in Cuba, and had been curator of the botanical gardens and herbarium at a museum in Pará, Brazil. Attracted by Baker's reputation, Wilson went to Pomona for one year as a special student. Baker was an energetic, inspirational teacher who instilled an extraordinary enthusiasm in his small group of students; he stressed

Wilson Popenoe standing beside a papaya tree in a California nursery, 1909. (*Courtesy of Hunt Institute for Botanical Documentation, Carnegie Mellon University, Pittsburgh, Pa.*)

hard work with a zeal that was almost holy. Encouraged by Baker, Wilson wrote his first horticultural articles in the *Pomona College Journal of Economic Botany.*

Looking back to 1910, Wilson Popenoe described many years later in some unpublished autobiographical notes how he first became interested in plant exploration:

> It was at just about this same time that I began to read of the great plant hunters. I thrilled at the tale of Lieutenant Bligh and his voyage in the *Bounty,* to bring the breadfruit tree from Tahiti to the British possessions in the West Indies: I pored over the story of Hasskarl, Markham, and Spruce, who combed the Andes for superior varieties of the quinine-yielding cinchona trees, and finally succeeded in establishing this precious drug plant in the Far East, where not only has it saved countless human lives, but has formed the basis of a highly profitable agricultural industry; and I followed the trails of Ernest H. Wilson and Frank N. Meyer, who were at that very moment exploring the remote parts of China, whence

Wilson Popenoe in a California nursery, 1911. (*Courtesy of Hunt Institute for Botanical Documentation, Carnegie Mellon University, Pittsburgh, Pa.*)

they were sending to the United States many ornamental shrubs and trees, and not a few economic plants which have in the past quarter of a century found congenial homes in our country.

All this inspired me, and I began to feel that plant hunting was just about the most romantic occupation imaginable. Not only did a chap get to travel in out-of-the-way corners of the world, but he stood a good chance of bringing home some new fruit, or food plant, which would add materially to his country's wealth and happiness. After all, the march of empire had gone hand in hand with the transplantation of crop plants from one part of the world to another. What would have been the history of the Malayan archipelago without rubber trees from the Amazon valley? How would Cuba have prospered without sugar cane from the Old World? What would Brazil, and Colombia, and Guatemala, and Salvador have done without the coffee plant? True enough, they would have got along somehow, but there would have been a very different story to tell.

And so I continued to make friends with plants. Behind our home in Altadena there was a small frame building familiarly referred to as "the shack," in which we boys slept; and in front of the shack was a small plot of ground which I took over for gardening purposes. In our neighborhood

> were the homes of numerous people from the East, people of means, who had come West to retire. Nearly all of them went in for gardening. I soon discovered that I could make pin money by supplying them with rose bushes, which were easily grown from cuttings in my little plot of ground. After school hours I would dig a dozen Maman Cochets, or Papa Gontiers, or Ulrich Brunners, and deliver them in a wheelbarrow to one of these people, collecting a dollar and a half in hard cash—just about the easiest money, it seemed to me, that any one could hope to earn.

Wilson's father, the versatile Fred Popenoe, had always been interested in gardening. Having lost most of his fortune and health in Costa Rica, he decided to start afresh in 1908 in a new business—a commercial nursery on his property in Altadena, which he called West India Gardens. He propagated and sold tropical and subtropical plants, ably assisted by his son Wilson and some other boys. During 1909, 1910, and 1911, Wilson traveled all over southern California looking for promising avocado seedlings worth propagating, frequently accompanied by a schoolmate and friend, Knowles Ryerson, who later was chief of the Bureau of Plant Industry of the United States Department of Agriculture (U.S.D.A.). Cash was scarce, so the boys covered a lot of territory on bicycles and on foot.

There was no plant quarantine in those days, and the avocado was growing in popularity. Hotels in Los Angeles used to import fresh avocados from Mexico for salads. The West India Gardens specialized in the avocado, but also sold some new and rare fruits such as the feijoa from Uruguay; the cherimoya from Mexico, Ecuador, and Peru; the white sapote from Mexico; and the loquat from Japan. At the time, there was a tremendous variation among the avocado seedlings growing in California. No two seedling trees produced fruit exactly alike, so it was necessary to propagate the best avocados asexually by grafting or budding, to obtain orchards of trees that would produce fruit of uniform quality. Mexico had been growing avocados for centuries, so Fred Popenoe decided in 1911 to send a young agricultural student named Carl Schmidt to that country to look for high-yielding and hardy avocados, suitable for the climate of southern California. Schmidt sent back budwood from forty-one selected avocado seedling trees from various regions of Mexico. As soon as this Mexican budwood arrived in the Altadena nursery, Wilson propagated it by budding on rootstock seedlings. He also photographed the fruit specimens and made pomological descriptions of them. One of Schmidt's selections was a variety named Fuerte from Atlixco in the state of Puebla, with a superb, creamy taste and tolerance to cold. The Fuerte, the

THE AVOCADO

A Salad Fruit of the Tropics

There is no more delicious fruit than the Avocado (Persea gratissima), or Alligator Pear, as it is commonly but incorrectly called. It is a tropical fruit, pear shaped or round, weighing from a half pound to a pound and a half or even more, green or purple in color, with a thick edible flesh surrounding a hard seed the size and shape of a small peach. The flesh, in color a rich golden yellow, is eaten raw in a number of ways, its rich, nutty, buttery flavor being very agreeable.

The Avocado offers great possibilities for culture in Southern California. It always brings a good price, single fruits selling as high as seventy-five cents each. It stands shipment well. The demand, especially in the large cities, has always been in excess of the supply.

The seedling trees grow to an immense size and are shapely and beautiful. They are prolific bearers, and hardy south of Santa Barbara. The natural season of fruiting is from September to November, but the fine budded varieties extend the season from July to April. The budded trees do not grow so large, being about the size of large, vigorous orange trees. The trees are propagated from seeds, and can be budded to whatever variety desired, although the seedlings themselves are in general use. From a sedling tree at Hollywood thirteen years old, the fruits in a single year, about 2000 in number, sold for as much as $325.

It is recommended, however, to purchase budded trees yielding a superior quality of fruit. Set budded trees the same distance apart as orange trees, or if space is available at a somewhat greater distance. Weekly irrigation and cultivation is necessary the first year; thereafter semi-monthly. Apply a reasonable amount of fertilizer.

No more profitable tree can be grown in Southern California than the Avocado, and the increasing demand and high prices will enable those who plant them in commercial quantities to reap a fortune.

Seedling Avocados.—We handle seedlings grown only from a superior quality of fruits. The flavor is first class, the tree a rapid grower, usually beginning to bear the fourth or fifth year from the seed, and is desirable as an ornamental tree, as well as for the fruit. From one to two feet tall, in gallon cans, each $1.00.

Budded Avocados.—Our stock of budded trees will not be ready before midsummer. Deliveries can be made in quantity and quality not excelled by any other grower, as we have procured our buds from the finest varieties. Price, balled, or in cans, each $3.00.

West India Gardens

F. O. Popenoe F. W. Popenoe

SUBTROPICAL PLANTS AND TREES

Santa Anita and Calaveras Streets

ALTADENA, CALIFORNIA

Home Telephone 401 (Pasadena Exchange)

An advertisement of Popenoe's avocado nursery, 1911. (*Courtesy of Hugh Popenoe.*)

Fred Popenoe at his nursery in Altadena about 1914. This two-year-old Puebla avocado tree is producing its first crop. (*Photograph by Wilson Popenoe; courtesy of Hunt Institute for Botanical Documentation, Carnegie Mellon University, Pittsburgh, Pa.*)

Spanish name for strong or robust chosen by Fred Popenoe for his best variety, turned out to be the world's leading commercial variety of avocado for many years. By 1938 it was estimated that over a million Fuerte trees were growing in California alone. Another important avocado variety found by Schmidt was the Puebla.

Unfortunately, despite the success of the Fuerte avocado, the West India Gardens lost money on its introduction. The cost of Carl Schmidt's explorations in Mexico was greater than the profits from the sale of the Fuerte trees. Nevertheless, Wilson's father had performed a very useful service for fruit culture in California.

Fred Popenoe next turned his attention to date palms. There was a growing interest in planting this crop in the dry interior valleys of southern California and Arizona, which he referred to as the "American Sahara." Planting material of good varieties was scarce. Since high-yielding varieties would not "come true" from seed it was necessary to propagate the date palms asexually, by planting offshoots or suckers from selected trees. One usually had to wait seven or eight years for these offshoots to attain the proper size for propagation—

Wilson Popenoe on an elephant, Saharanpur, India, on the way to Middle East, September 1912. (*Courtesy of American Heritage Center, University of Wyoming.*)

up to about forty pounds in weight. There were few choice varieties growing in the United States, so any date palm planting on a commercial scale would require the importation of selected offshoots from Arabia, Persia, Iraq, and North Africa. There was an additional problem: date palms are unisexual; some are male and others are female. Therefore, it would be necessary to obtain offshoots of both sexes.

Fred Popenoe managed to obtain orders for several thousand date palm offshoots from the United States Department of Agriculture as well as from some enterprising agriculturists in the Imperial and Coachella valleys of California. So in July 1912, having borrowed some money, he decided to send Wilson, who was twenty, and his older brother, Paul, who was twenty-three, for a year to the date-producing countries of the Old World to obtain a large number of offshoots. Paul had already spent a few months in Algeria the year before, and knew some Arabic. Wilson and Paul then set out for the Orient, via San Francisco, Hawaii, Japan, and on to Calcutta. In his unpublished reminiscences, Wilson vividly described their experiences:

> Northern India gave me an opportunity to familiarize myself with mangos in the world's greatest mango-growing region; but it gave us some

The date palm, *Phoenix dactylifera*. (*From M. E. Descourtilz,* Flore Pittoresque et Médicale des Antilles, *2d ed. 1833. Courtesy of Botany Libraries, Harvard University, Cambridge, Mass.*)

Basra, Iraq, February 1913. Paul Popenoe, Wilson's brother, with some of the seventeen thousand date palm offshoots they shipped to California. (*Courtesy of American Heritage Center, University of Wyoming.*)

hard jolts, also, for it was here that Paul contracted the typhoid fever which nearly cost him his life a few weeks later at Basra, in the Persian Gulf; and it was here that I contracted the malaria and dysentery which laid me up for several weeks in the little British hospital at Maskat on the Arabian coast.

While I was getting my forty grains of quinine daily via the hypodermic needle, Paul, accompanied by Homer Brett, American consul at Maskat, made a journey inland to Wadi Samail, home of the Fardh date, one of the world's great commercial varieties. The party returned hastily, after having reached its objective: twice they had been fired on by Arabs. Since they had been under escort—the sultan of Oman having sent camel-troopers with the American consul—the matter was not to be treated lightly. A detachment was promptly dispatched by the sultan to punish the offenders, and a petty war was started which lasted intermittently for more than a year.

At Basra our real work began. After Paul's recovery from typhoid (he probably would not have recovered at all, had he not been taken in by the Cantines, American missionaries of the Dutch Reformed Church), we bought several thousand young date palms along the banks of the Shatt-el-Arab, and began preparing them for shipment. While the work continued, we went up to Baghdad on a river steamer and bought several thousand palms in that region. . . .

We suffered from the cold. Though burning hot in the summer months,

In 1912–1913 Wilson Popenoe and his older brother, Paul, made a long trip to the Middle East to obtain propagating material of the date palm to plant in California. (*From Paul Popenoe,* Date Growing in the Old World and New, *Altadena, Calif.: West India Gardens, 1913.*)

FIFTY-THREE POUNDS OF DATES

the Persian Gulf country is bleak and disagreeable in the winter. After having been shot at once more, we experienced something akin to a sense of relief when our nine thousand palms were safely stowed on board the *Mokta,* a tramp steamer from Swansea, and headed down the Persian Gulf. In the Red Sea we lost quite a few of them: they were on deck, and were washed overboard in a storm. Then there was a long dry spell and the ship was short of drinking water. [Wilson was obliged to make a trade: he had to give the captain his typewriter—the only thing of value he owned—so the captain would make water available for the remaining date palms, which otherwise might have died.] After forty-nine days of the poorest eating I have ever experienced, we landed at Oran, Algeria, and went down to Biskra in the Sahara, where Paul made arrangements for the purchase of several thousand more offshoots and left me to carry on the work of packing them, while he went to London to arrange transshipment of our Arabian palms, which finally landed at Galveston and went overland in seventeen freight cars to California.

Algeria, for comfort, was a marked contrast to Arabia; and it was picturesque enough, though one did not have the feeling of being far from European civilization, as one did in Baghdad. My stay was enjoyable though short. Our funds also were short, by this time, and when Paul telegraphed from London, asking me to meet him in Algiers, I had to dispatch Muhammad ben Ali across the desert some thirty miles to borrow twenty francs from his brother Abderraham (our agent) so that I could buy a railway ticket.

PACKING DATES IN THE SAHARA

Conditions were unsanitary in this region. In 1912 Paul Popenoe almost died of typhoid fever, and Wilson suffered from malaria and dysentery. (*From Paul Popenoe,* Date Growing in the Old World and New, *Altadena, Calif.: West India Gardens, 1913.*)

> I left Paul to complete the job, and sailed for New York. I was anxious to get back to the States for two reasons: first, we had been gone nearly a year, which was a long time for a youngster getting his first taste of foreign travel (and what a baptism of fire it had been!); and second, I had been in correspondence with David Fairchild of the United States Department of Agriculture, who had offered to appoint me as an agricultural explorer the moment I reached Washington. . . .
>
> The fascination of plant hunting was upon me, stronger than ever; though sometimes at night I dreamt of those days in Basra when Paul lay between life and death, and even now I can hear the shrill cries of the sentinels on the forts at Maskat, as they signalled each other during those long nights that I tossed with fever on my bed in the little hospital. Eighteen hundred dollars a year looked like a fortune to me; I gave up all thought of finishing my college work and embarked upon a series of voyages which were to take me to many remote and interesting corners of tropical America.

The imported date palm offshoots grew well in California. Fred Popenoe himself bought a farm near Thermal in the Coachella Valley, subject to a mortgage, and developed a promising date palm plantation. However, just as the palms came into bearing, the price of dates dropped sharply in the U.S. market. Fred lost the farm since he was unable to meet the mortgage payments. This was unfortunate, to say the least: this date palm plantation set out by Fred Popenoe was at the time the largest in the United States.

In 1919 Fred Popenoe sold his nursery at Altadena and retired from active business. He then established an orchard of avocados and plum-like loquats at Vista, California, where there was a severe water shortage. Virtually singlehanded, he organized and promoted the Vista Irrigation District, which provided the northern end of San Diego County with an abundant water supply. Fred was one of the organizers and president (in 1922) of the California Avocado Association. When Fred Popenoe died in 1934, his obituaries concurred that he would be remembered especially for his remarkable introduction of the Fuerte avocado in 1911, his date palm introduction of 1913, and his constructive work at the Vista Irrigation District toward the end of his life. It should also be noted that several young men, including his son Wilson, were taught the rudiments of subtropical nursery work at West India Gardens.

The day after Wilson Popenoe returned to the United States in 1913 from the date palm expedition, he had to make an important decision: he was twenty-one years old and had been offered a four-year scholarship at Cornell. He went to Washington and called on David Fairchild, chief of the Office of Foreign Seed and Plant Introduction of the Bureau of Plant Industry of the United States Department of Agriculture, seeking his advice. Fairchild talked Wilson out of going to Cornell and offered him instead the job mentioned above as an agricultural explorer for the U.S.D.A. at $150 per month, which Wilson accepted. Upon hearing the news, his father wrote him he should be ashamed to soak the American people so much. Many years later, David Fairchild recalled the occasion. In an interview with a reporter from the *Miami Herald* in 1950, he stated, presumably with tongue in cheek: "We were afraid college would ruin Wilson Popenoe. We wanted to make a plant explorer of him, and were afraid that he wouldn't be much good for the U.S.D.A. in the field after he got filled on book learning."

Chapter II

Brazil, 1913–1914: The Navel Orange

Following a few months in Washington learning the routine of the United States Department of Agriculture, and a brief visit to Florida to study mango growing, Wilson Popenoe sailed for Brazil on October 5, 1913, on the SS *Vandyck,* to investigate the history and cultivation of the navel orange. He was the junior member of a party of three, which included two departmental veterans, A. D. Shamel and P. H. Dorsett. A few days earlier, David Fairchild, in charge of the Office of Foreign Seed and Plant Introduction, had called the force together in the clerical room of his office in Washington. After a short talk on the subject of the forthcoming expedition to Brazil, he called on its members for a few personal remarks. Each explorer responded in a way that seemed to indicate appreciation of the importance of the work and the responsibilities it entailed. The group then adjourned to the roof of the building where the official photographer took their picture.

By chance, ex-President Theodore Roosevelt, his son Kermit, and several friends, including a well-known tropical explorer, George K. Cherrie, were passengers on the *Vandyck.* From early childhood, Wilson had been an ardent admirer of Teddy Roosevelt, whom he considered the greatest American of his time. Strolling on the deck, he had some long conversations with Roosevelt, who was interested in foreign plant introductions from the tropics. He pointed out to Wilson, however, that we should do more to cultivate and improve our own native plants and fruits, such as the red, fleshy, edible buffalo berry, from shrubs growing in the western United States.

Wilson had hoped to sit at a table with some Brazilians, so he could practice his Portuguese, but to his disappointment he was seated with five other North Americans. He was, however, delighted by his introduction to the West Indies. Since Theodore Roosevelt was on board, the captain made a special effort to allow the passengers to get a close view of St. Kitts, Nevis, Antigua, Guadeloupe, Dominica, and Martinique. Wilson was much impressed by the splendid views, by the

Employees of the U.S.D.A. Office of Foreign Seed and Plant Introduction, September 30, 1913. David Fairchild, chief, on the right. The three plant explorers, about to leave for Brazil, kneeling, left to right: Wilson Popenoe, A. D. Shamel, and P. H. Dorsett. (*Courtesy of the National Archives, Washington, D.C.*)

natural beauty of the islands and the Caribbean "emerald green, turquoise blue, or indigo, accordingly as the water is shallow or deep." Going ashore at Barbados, he noted the tropical flora of that island did not have the luxuriance of growth he had seen at Singapore and Penang. The thin soil on coral rock seemed in places to stunt the growth of the trees. The native population appeared to be happier, the streets cleaner, than in the Far East. On the whole, Wilson found the combination of beautiful and healthful surroundings more attractive than the Far East, where many charming sights were marred by being surrounded by disagreeable people and offensive smells. The streets in picturesque Bridgetown, for example, were not used as open sewers as was so often the case in the Far East. Agriculture, including cane and cotton, was highly developed in Barbados. Coconuts, breadfruit, tamarinds, corn, cassava, yams, and pigeon peas were widely planted. Travelers palms, cabbage palms, colorful frangipani shrubs with star-shaped yellow flowers, and crimson hibiscus plants were to be seen all over the island.

The *Vandyck* proceeded to Bahia, the first port of call in Brazil,

On board the SS *Vandyck* en route to Brazil, October 5, 1913. This was Wilson Popenoe's first trip as an agricultural explorer for the U.S.D.A. Left to right: P. H. Dorsett, A. D. Shamel, Wilson Popenoe. (*Courtesy of American Heritage Center, University of Wyoming.*)

anchoring quite a distance off the shore. Several small vessels came out in style to greet Theodore Roosevelt, their decks crowded with public officials resplendent in frock coats and top hats. Roosevelt was on his way to explore the River of Doubt in west-central Brazil where he nearly perished. Years later, George Cherrie told Wilson how the ex-president and his small party had battled their way for many days through the rapids of the River of Doubt. Their canoes were severely damaged and had to be abandoned. The group struggled on foot in the impenetrable jungle, where Roosevelt suffered a leg injury so severe he could barely walk. One night he said, "Boys, some of us are not going to get out of here alive. Cherrie, I want you and Kermit to go on. I will stay here." Kermit and Cherrie begged for one more day's effort. How could they abandon Theodore Roosevelt to die a lonely death in the Brazilian wilderness? Just when the situation seemed hopeless, they found a way out and the ex-president survived. Wilson Popenoe was deeply touched by this story and by Cherrie's devotion to Theodore Roosevelt.

In Bahia, Wilson and his U.S.D.A. companions hastened ashore to

The orange, *Citrus sinensis*. (*From M. E. Descourtilz,* Flore Pittoresque et Médicale des Antilles, *2d ed. 1833. Courtesy of Botany Libraries, Harvard University, Cambridge, Mass.*)

catch their first glance of the navel orange in its native environment. In his unpublished notes, Wilson described the origin and history of this fruit:

> Among the many plants brought to Brazil by the Portuguese in colonial times was an orange known as Selecta. This is a fine large orange, said to have reached Brazil from the Azores or one of the other islands off the coast of West Africa. It is still cultivated in Brazil, particularly in the vicinity of Rio de Janeiro, where the fruit is commonly seen in the market.
>
> Selecta is an orange containing numerous seeds. To a greater degree than other varieties, it has the habit of throwing "bud sports": that is, occasional fruits with more or less well-defined navels will be found among the normal fruits. In cases where these navels are strongly developed there is a tendency for seeds to be few in number and, occasionally, for them to be missing entirely.
>
> Sometime in the very early part of the nineteenth century, tradition says, a Portuguese gardener whose name has been lost, noticed the occurrence of these "bud sports" on the Selecta orange, and taking one of the twigs which produced navel oranges free from seeds, he grafted it onto a young sour orange tree. When the grafted plant bore fruit, it yielded nothing but seedless navel oranges. The obvious superiority of this variety

> resulted in its further propagation, and it was not long until there were small groves around Bahia, where the variety began to be called "laranja de umbigo," or navel orange.
>
> Shortly after the middle of the nineteenth century—when this variety was still relatively new, but had nevertheless come to be well-known at Bahia and highly esteemed by all who were familiar with it—the Reverend F. I. C. Schneider, an American missionary who was living at Bahia, conceived the idea of sending trees of this variety to his native land, where no seedless oranges were grown at that time.
>
> Arrangements were made with William Saunders, who was in charge of the gardens and grounds of the young Department of Agriculture at Washington, to receive grafted plants of the navel orange from Bahia. A shipment was made, but resulted in failure. A second lot of plants was forwarded—this in 1870—most of which survived. In the greenhouses at Washington they were multiplied by Mr. Saunders, who sent young trees to California and Florida. The variety was never commercially successful in Florida; but in California, two trees sent to Mrs. Eliza Tibbetts of Riverside began to attract attention as soon as they came into bearing, and it was not many years before nurserymen took up the "Washington Navel Orange," as it was called, and put in on the market.
>
> The growth of the navel orange industry of California, from its inception in the 1870s to the present day, has been an outstanding example of what can be achieved when horticultural science is linked with cooperative effort. From every point of view it has been a remarkable job; but when we are considering its importance today, we should not forget what we owe to the unknown Portuguese gardener in Bahia, who first realized the value of this fruit and propagated it; to the Reverend Schneider, who wanted his countrymen back in the States to enjoy seedless oranges just as he was doing in Bahia; to William Saunders, who nursed and multiplied the first sickly plants which came from Brazil, then a long way from the United States in point of time; and to Mrs. Tibbetts, who planted and cultivated the trees which demonstrated the possibilities of the navel orange in California.

Although only in Bahia for a day before going on with the *Vandyck* to Rio de Janeiro, the U.S.D.A. plant explorers found the time to visit nearby Cabullo. There, on October 20, 1913, they observed the two navel orange trees they had come thousands of miles to see. These trees, which forty years earlier had given birth to the immense navel orange industry of California, were flourishing in their native home, which at first glance appeared to be a fine-looking orchard. Upon closer inspection, however, Mr. Shamel discovered mottled leaf disease caused by the action of nematodes on the roots, and wither-tip fungus on several fruits. These were also common problems in California. The soil was evidently rich, a deep red clay quite free from stones.

Upon returning to the ship, the plant explorers received a promise of full cooperation from three officials of the Brazilian Department of Agriculture. Wilson had shown the Brazilians a letter of introduction from Dr. B. T. Galloway, assistant secretary of agriculture of the United States, which stated the following:

> F. W. Popenoe has had special training in subtropical horticulture, having published in encyclopedias of horticulture on such subjects, having been trained by a botanist who spent several years in Brazil and having himself made an exploring trip to India and Arabia at his own expense. He is a young man and was employed expressly for exploration work and requires only the training which this trip to Brazil with older men will give him to fully fit him as a permanent agricultural explorer for the United States Department of Agriculture.

Several days later the plant explorers arrived in Rio de Janeiro and unanimously agreed it was the finest city they had ever seen—so cosmopolitan and European in character it didn't seem like Latin America. In 1913 there was little traffic (and no pollution as yet) on the Avenida Central, a broad, clean street lined on each side by handsome buildings and divided down the center by a row of ornamental trees. The sidewalks were wide, uncongested, and protected by luxuriant shade trees. The cafés were typically French and reminded Wilson of Algiers. Many types of *refresco* were available, and he sampled them all, including pineapple, papaya, Surinam cherry, coconut, orange, lemon, and many other minor fruits. The city was full of striking avenues adorned by huge cabbage palms (*Roystonea oleracea*); the Avenida de Mangue was impressively lined with four rows of these very tall trees.

Wilson was obliged to buy a white duck vest with pin stripes. He was informed respectable men were expected to wear at all times not only a coat but also a stiff collar, tie, and vest when walking about the streets of Brazilian towns, regardless of the temperature.

Wilson, Mr. Shamel, and the American ambassador, Mr. Morgan, paid a visit to His Excellency the Minister of Agriculture, Industry, and Commerce, the Honorable Dr. Pedro de Toledo. They were ushered into a large, lavishly furnished reception hall, considerably finer than anything of the kind at the United States Department of Agriculture back in Washington. The Brazilian minister expressed interest in the North American mission and indicated he was anxious to cooperate in every way possible. Upon leaving, they all shook hands several times around with great formality, and Wilson made a fine debut in diplomatic contact with Latin American government officials—a task at which he excelled.

Wilson visited the Jardim Botanico (Botanic Garden). Its new director, an Englishman named Dr. Willis, was attempting to improve the conditions of the garden, which was a bit run down. The outstanding feature of the garden was the magnificent avenue of tall cabbage palms, which impressed Wilson as the most superb planting of any kind he had ever seen. These immense palms, like many others in Rio, had been grown from seeds of a tree imported by Dom João VI in 1809; in many cases, their trunks had a splendid copper hue.

Another impressive avenue of the Jardim was planted with the champac tree of India, *Michelia champaca,* sacred and dedicated to Vishnu in its native land. The champac, a splendid street tree, tall and shapely with beautiful light green leaves, reminded Wilson of the avocado. There was also a striking avenue planted with mangos, which had been imported a century earlier from Goa. The mango trunks had been profusely hacked in order to produce fruiting, but to no avail. Numerous guava varieties were to be seen as well as many other evergreen trees and shrubs of the myrtle family. All in all, Wilson believed that if sufficient government funds could be made available, the Jardim Botanico could become one of the finest botanic gardens in the world, having such a magnificent nucleus of palms.

Wilson and Messrs. Dorsett and Shamel then returned to Bahia from Rio on the Royal Mail Steam Packet ten-thousand-ton ship *Alcala.* Wilson seemed to prefer Brazilian cooking to the monotonous British fare of porridge and beefsteak. Back in Bahia they were provided with a Brazilian interpreter, Snr. Argollo, who was also an agricultural expert. By this time, Wilson himself could carry on a conversation in Portuguese, although his companions could not. In point of fact, they were jealous of Wilson's linguistic talent.

The plant explorers visited several splendid orange orchards in the vicinity of Bahia. The trees were vigorous for the most part and carrying a fine crop of fruit. The deep, rich reddish clay seemed to be an ideal soil for orange growing: it reminded Wilson of the soil at Redlands, California, while the tropical vegetation reminded him of Costa Rica, with a profusion of bananas, palms, mangos, tamarinds, jackfruit, and breadfruit. His favorite flowering plant was widely existing: the frangipani, *Plumeria rubra,* with its coral and creamy pink flowers borne in great clusters at the end of the branches. An immense number of evergreen cashew nut trees, *Anacardium occidentale,* were to be seen. Although vigorous, numerous stems were gnarled with

In 1913, Wilson Popenoe, as an agricultural explorer for the U.S.D.A., collected jaboticaba seeds. These fruits, produced directly on the tree's trunk and larger limbs, are consumed fresh in southern Brazil and also used to make jellies and wine. (*Photograph by Wilson Popenoe.*)

crooked branches. Northeastern Brazil, near the equator, is probably the region where the cashew originated.

Snr. Argollo invited Wilson and the others to his home on the outskirts of Bahia. An elaborate midday meal was served, starting with *vatapá,* a popular local, yellowish, highly seasoned mush. Its most important ingredient was the oil of the dende palm (African oil palm), *Elaeis guineensis.* Many years later, Wilson was to become involved in the cultivation of this palm for the United Fruit Company at Lancetilla, Honduras. The currylike vatapá also included fresh fish, dried shrimps, peanuts, and cassava flour. Cassava, also called manioc, mandioc, or yuca, is a most important tropical root crop (*Manihot esculenta*), from which a flourlike meal known as *farinha* is prepared.

The vatapá, which was the first course, was followed by a second course of baked fish; a third course of baked tongue with rice and salad; a fourth course of roast turkey with farinha stuffing; then, a fifth course of roast beef, followed by a large platter of local fruits, cakes, and coffee. By this time, the plant explorers were overwhelmed.

Bahia, Brazil, November 1913. P. H. Dorsett and Wilson Popenoe preparing jaboticaba seeds for shipment to Washington. (*Courtesy of the National Archives, Washington, D.C.*)

During the next two months they surveyed the terrain around Bahia, collecting plant material and herbarium specimens, which they shipped to the U.S.D.A. in Washington. For the most part the explorers, while in Brazil, did not cut themselves off from civilization, nor blaze a path in the wilderness. Wilson pointed out in his notes:

> An agricultural explorer, it seems to me, can, as a general rule, obtain more material of value in a cultivated region than he can by penetrating into the wilds. Agricultural crops accompany man, and are not found in the untracked wilderness. Considering most of the important food crops introduced into the United States, I believe they will be found as a rule to have come from a region where their value was realized and their culture carried on extensively.

They studied a planting of jaboticaba fruit trees (*Myrciaria cauliflora*), which have the peculiar habit of producing their fruit directly upon the trunk and larger limbs—like the cacao or cocoa tree (*Theobroma cacao*). The jaboticaba, popular in Brazil, may be consumed

Wilson Popenoe in his hotel room in Lagoa Santa, Minas Geraes, Brazil, writing a report to the U.S.D.A., February 1914. (*Courtesy of American Heritage Center, University of Wyoming.*)

as a fresh fruit; it is also used to make wine, as well as tasty jellies. Its whitish, juicy pulp has a flavor strongly suggesting a good grape.

On the streets of Bahia, Wilson found the most delicious pineapples he had ever eaten, known locally as *abacaxi.* These large pineapples were not cultivated around the city, but across the bay, and were brought to Bahia in small sailing vessels, then peddled about town in large baskets fastened across the backs of mules. He observed: "I am becoming more and more of the opinion that the pineapple must be classed as the finest of the tropical fruits."

In January 1914, while Mr. Shamel returned to California, Wilson and Mr. Dorsett departed for an agricultural exploration of the interior of Brazil. They returned to Rio, then traveled through Minas Geraes to the Rio São Francisco, whence they proceeded downstream by river steamer to Joazeiro, from which town they took a train back to Bahia.

In this little-known back country they hoped to find crop plants that could be of value in the southern United States. In Minas Geraes they studied a fruit-bearing shrub called pitanga (*Eugenia uniflora*), also known as the Surinam cherry. Its highly aromatic fruits, which look like tiny ribbed tomatoes, were used in Brazilian cafés for sherbets—and also for making excellent jams and jellies.

They passed through the cattle-, corn-, and coffee-growing region of southern Minas Geraes, and visited the agricultural college at Lavras which had been developed by North Americans. This school gave a three-year course along general lines, followed by three years of agricultural study. At the time, there were students from all over Brazil, including about 120 men and a few women.

Their trip down the Rio São Francisco started by launch from Pirapora at the head of navigation of the river. Some hours later they arrived at a ramshackle village on the river bank, where they boarded a small paddle-wheeled steamer. Mr. Dorsett took numerous photographs as he did elsewhere in Brazil. As they proceeded downstream, the population became more and more sparse. Once in a while they passed a primitive hut made of poles chinked with mud and thatched with grass. There were scattered plantings of corn, cassava, bananas, and cane. Two hundred miles below Pirapora lies the town of Januaria, deep in the cattle country, where lace and knives were also produced. It was the custom in Januaria to allow goats, sheep, and hogs to live right in the houses, like pets, instead of keeping them outside. Sheep were being used as draft animals instead of oxen.

The most common fruit in this region was the imbu, or hog plum (*Spondias tuberosa*), which was eaten fresh, but also used to prepare the dessert called *imbuzada.* The pulp of the imbu was passed through a strainer and sweetened; then curdled milk was added to form a delectable dessert with the consistency of a cream soup. The imbu tree itself furnished dense shade under which cattle gathered during the heat of the day.

Boarding another larger steamer, which was carrying malodorous raw hides, they proceeded down the river to the mouth of the Rio Carinhanha. En route, they shared a cabin with twenty Brazilians. In Carinhanha they met a pessimistic local padre who traveled up and down the river in a small canoe to visit his flock. He bewailed the deplorable, miserable condition of the region and the lazy people who lived there. According to him, these loafers refused to work because they were quite content with a few fish they might catch, a little cassava, corn, salt, and tobacco, and were beholden to nobody. He considered it to be a backward region just as it must have been in the era of Adam and Eve, with little hope for the future.

One of the stops on the river was at Pilão Arcado, where wax was produced from the carnauba palm (*Copernicia prunifera*). Carnauba wax was obtained by cutting the leaves and drying them in the sun; after several hours the wax exuded in the form of a powder used in making candles, polishes, varnishes, lipsticks, and phonograph records. The hard, brittle candles burned with a clear, yellow light. The fruit of the palm was used for hog feed.

The diet on the steamer consisted of a monotonous round of beef, rice, and beans. Wilson regretted he had not brought a filter along; at times they had to drink water directly from the river, but apparently suffered no ill effects. Near the final destination on the river at Joazeiro, the mosquitos became bothersome at night, but there was no room in the crowded cabin to rig up a mosquito net. Citronella seemed to help a little, however.

After the poverty-stricken, backward villages on the Rio São Francisco, Joazeiro, with a population of twelve thousand, seemed a prosperous, bustling metropolis, with its railroad and many vessels of every description loading and unloading cargo. This port was some 850 miles down the river from Pirapora, a trip of about ten days.

On the railroad trip from Joazeiro to Bahia they stopped for two days at Bom Fim to collect two sacks full of imbu seeds, as well as other plant material. In early March of 1914 they returned to familiar surroundings and the comforts of civilization at Bahia. A month later they sailed back to the United States on the SS *Vauban,* taking with them a number of Brazilian products including cashew wine, jaboticaba juice and jelly, cashew nuts, jackfruits, guava jelly, imbu jelly, orange budwood, Angola grass seed, palm seeds, and cactus seeds, as well as many other seeds.

While in Brazil, Wilson had the opportunity to read Mark Twain's *Innocents at Home.* Twain asserted the tropics were overrated: no tropical land with a monotonous, unvarying climate could be beautiful; change was the handmaiden nature required to do her miracles with. By contrast, temperate zones with their four well-defined seasons created a world of beauty and the enjoyment of watching the gradual, harmonious development of each season.

Passing through the West Indies on the way back to the United States, Wilson came to the conclusion Twain was mistaken. Wilson did not mind at all the heat he had felt at times in Brazil—in fact he enjoyed it. As to beauty, he was of the opinion there was no region on earth other than the tropics where such luxury of vegetation and magnificent forms of plant life could be found. He stated in his notes: "I believe if Mark could have seen the palm avenue in the Jardim Botanico at Rio de Janeiro, he would have been forced to admit that the North could not produce anything to surpass it."

Wilson and Dorsett returned to New York harbor on April 12, 1914. A letter from their chief, David Fairchild, was awaiting them, written from his home "In the Woods," Chevy Chase, Maryland.

> To the Explorers returning: I am mighty glad to feel that you are safe again home. We certainly missed you both and have watched your progress with keen interest. . . . The thing that will surprise you both is that without either of you we have gone ahead and done things. It shows at last we have a machine with enough good people in it to run on for many years after we are all dead and buried. . . . Dorsett's room here is in readiness, but Popenoe's, poor fellow, has been requisitioned for a quarantine station. He will find a nook though. We are all looking forward to your return, and all feel that you have done a fine piece of exploration work.
>
> Welcome home,
> Yours as ever,
> David Fairchild

On April 16, Fairchild gave a dinner at Freund's Restaurant in Washington in honor of Dorsett and Wilson Popenoe. The chief read a long and witty poem he himself had written entitled "The Quest of the Jaboticaba." Here are a few of the verses:

> The men have returned, they are back from Brazil,
> And the stories they tell would make your blood chill.
>
> T'was a weird kind of fruit that they went for you see,
> Growing all up and down on a slow-growing tree.
>
> Few men who can do what they surely have done
> Caught the Jaboticaba without using a gun.
>
> No wonder they reach for it, eager to see
> What a new kind of drinklet Jaboticaba could be.
>
> We welcome them back from the jungles and streams,
> Of Brazil the mysterious, the land full of dreams.

Chapter III

Florida and Cuba, 1915: The Mango

Back in Washington, Wilson spent some time working on a U.S.D.A. bulletin describing his Brazilian expedition; he also visited his parents in California. His boss, Dr. David Fairchild, whom he called "Chief," was indeed the chief agricultural explorer for the U.S.D.A., and enthusiastic about the possibilities of growing mangos, *Mangifera indica,* in southern Florida. Selected mango varieties from many tropical countries were on trial at the Plant Introduction Garden in Miami. Season after season, the vigorous evergreen trees would grow well but produce virtually no fruits—perhaps fewer than a dozen mangos per tree.

So in mid-1914, Dr. Fairchild told Wilson to prepare himself for a lengthy visit to India, the aboriginal home of the mango, to find out what was wrong with the Florida mangos. Native to south Asia, eastern India, and the Malay Archipelago, and prominent in Hindu mythology since ancient times, the mango has been cultivated by man for over four thousand years. It is widely grown in all warm countries throughout the world.

Wilson was looking forward to his trip to India when World War I broke out in August 1914. A change of plans was necessary. The Chief told him to postpone the expedition to India until 1915; the war would probably be over by that time. Meanwhile, Wilson was to go to Florida and Cuba to learn more about mango growing.

A few days later, Wilson was in Miami at the little laboratory on Brickell Avenue, which Henry M. Flagler had donated to the U.S.D.A., Office of Plant Introduction. Immediately across the street was the six-acre Plant Introduction Garden, which in Wilson's opinion did more real good on a per-acre basis than any other agricultural station our government had ever operated. Unfortunately, it was wrecked by the hurricane of 1926, after it had been instrumental in making a success of the avocado and mango industries. A great number of ornamental plants had passed through its gates into the gardens and dooryards of Florida.

Wilson Popenoe under a Sandersha mango tree in plant introduction station, Miami, Florida, August 1913. The mango, *Mangifera indica,* a sacred tree in India, is a native of southern Asia. (*Courtesy of American Heritage Center, University of Wyoming.*)

The mango has been called "the apple of the tropics." The common seedling mangos of the West Indies, however, although juicy, handsome fruits, are sometimes so full of fiber the British suggest they should be eaten while sitting in a bathtub. In contrast, the fine, grafted mangos of the East, through many centuries of selection and improvement, have become superb fruits—as free from fiber as a peach with a pleasing, aromatic taste reminiscent of the nectarine, peach, and apricot. But in Florida, the best mango varieties failed to produce crops. Dr. Fairchild thought faulty pollination might be the cause; others believed the pollen of highly bred varieties was not potent enough to effect proper fertilization and consequent fruit production. Another hypothesis, which proved to be more accurate, was that the mango tree itself needed a rest from time to time to build up nutrient reserves so fruit buds might be formed. To check the growth of the mango trees some success was obtained by "ringing" (cutting away

The mango, *Mangifera indica.* (*From M. E. Descourtilz,* Flore Pittoresque et Médicale des Antilles, *2d ed. 1833. Courtesy of Botany Libraries, Harvard University, Cambridge, Mass.*)

bands of bark from the larger limbs). Throughout the tropics it was common practice to slash the trees annually with machetes, drive nails into them, or pound the bark with clubs. There was a popular West Indian refrain: "A woman, a mule, and a mango tree, the more you beat them, the better they be." Wilson and the other U.S.D.A. experts, rather than suggest such drastic procedures, limited their recommendations to the following program: (1) confine the commercial cultivation of fiberless mangos to the driest climates, avoiding rich, moist soils; and (2) choose the varieties with extreme care.

In 1915 Wilson visited Miami Beach with two female friends, crossing over a rickety, wooden bridge from the mainland in a bus. They saw a pavilion of sorts, a few modest houses, and an avocado grove on Collins Avenue over one hundred acres in size—said to be the largest avocado planting in the world at the time.

South Florida intrigued Wilson—it seemed to him like a bit of the West Indies tacked onto the southern part of the continental United States. Many years later he observed: "There is something in the air in Florida which literally smells of the tropics, something which my

Wilson Popenoe in Miami, 1914. (*Courtesy of Hunt Institute for Botanical Documentation, Carnegie Mellon University, Pittsburgh, Pa.*)

tropical soul has always missed in California." He stressed, however, that since south Florida rose out of the ocean, its framework was lime and sand, and its soil was poor. To grow plants well in Florida, one had to fertilize the soils and keep on fertilizing them. He mentioned the tragic example of the boom in Florida in the 1920s. Beautiful citrus and avocado groves which had been built up by years of fertilizing and cultural attention were abandoned when the craze for subdividing hit Miami. Within a few months the trees had a lean and hungry look; within a year they suffered badly and many of them died.

Dr. Fairchild thought the mango project would be furthered by having Wilson study the crop in Cuba. So in 1915 Wilson went to Havana, where North Americans were popular and times were good. It was the era of the *vaca gorda* or "fat cow," as the Cubans called it. The cafés were lively and merry, the theater popular. He witnessed several amusing Spanish zarzuelas—a sort of abbreviated musical comedy immensely popular at the time, which was destined to be replaced by the movies. Cuba had not yet gone through the agony of the collapse of the sugar market following World War I.

For several weeks Wilson toured the island from Pinar del Río to Santiago de Cuba, accompanied by a young Cuban agriculturist named Gonzalo Fortún. They saw many mango plantings while they admired the lovely *vegas* (alluvial flats) of the limestone region west of Havana, the beautiful Yumuri valley near Matanzas, and the verdant mountains of the Oriente. Stately royal palms and clumps of feathery bamboo graced the landscape. They went to a few cockfights and smoked a lot of strong black cigars. Wilson was an avid cigar smoker and connoisseur of fine tobacco, and in his opinion, Cuban cigars were the finest in the world. He pointed out good tobacco required more than good seed, favorable soils, proper culture, and careful curing. The mainland of tropical America in general from Brazil to Mexico did not produce high-class cigars: they lacked the "body" of Cuban tobaccos and fell short of the desired aroma. Just why this was so remained a mystery. He thought the cigars of Santo Domingo came closest to the high standard of Cuba.

Years later, poor quality tobacco continued to be a problem in Honduras. Samuel Zemurray, head of the United Fruit Company, called upon the president of Honduras. "General," he said, "what makes these Honduras cigars so rotten? Honduras tobacco used to have a great reputation in colonial times; it was appreciated clear down to Peru. I'm going to get you a Cuban who knows all about cigar tobaccos, and I'll pay his salary while he finds out what's the matter."

Wilson Popenoe, who was working at the time for the United Fruit Company in Honduras, was sent to Cuba to find the expert tobacco man. Through his old friend Gonzalo Fortún, by then director of an experiment station, the ideal Cuban, who had been "born under a tobacco plant," was selected for the job in Honduras. In the Copán tobacco region the Cuban built a curing shed of the most approved Cuban type; he planted the best Havana seed; he even went so far as to insist on having fiber from the royal palm in which to wrap his cured tobacco, like they did in Cuba, since it was supposed to add a certain delicate flavor to the finished product.

At the end of the first season a good crop was lifted. After four or five months the tobacco had been properly cured and the cigars were ready for testing. Wilson's old friend Shamel from the U.S.D.A., an inveterate cigar smoker who had been with him in Brazil years earlier, was asked to try the new cigars. He lit one and took a few leisurely puffs. Unable to control his curiosity, Wilson asked Shamel for his opinion. "I've smoked worse," was the reply, "though I can't recall just where." (To end this tobacco story on a happy note, it should be pointed out that following years of technical research and experimentation, very fine cigars are produced in Honduras today.)

After their mango-hunting trip to the eastern end of Cuba in 1915, Gonzalo Fortún and Wilson returned to Havana. Wilson was sufficiently *de confianza* to be taken into the bosom of the Fortún family. This was a singular honor for a North American gringo and enabled him to get an intimate glimpse of Cuban home life not often granted to the casual visitor. Latin Americans are traditionally slow to invite strangers to their homes, a trait that may come down to them from their conservative Iberian ancestors.

The Fortún house was a typical downtown Havana residence. The ground floor was devoted to business purposes, while the family lived on the second floor, reached by a handsome marble staircase which led off the street. Like most tropical residences, the rooms were high-ceilinged, amply ventilated, and devoid of carpets. The *sala,* or living room, was furnished with cane-seated chairs. There was a marble-topped center table, a gilt-framed mirror, a few pictures of European origin, and a piano. Gonzalo's pretty younger sister was a talented pianist who sang catchy, popular Cuban songs to Wilson as she flashed alluring glances to him over her shoulder—which thrilled him, he confessed, right down to his toes. Wilson's charm, good humor, and ability to speak fluent Spanish made him a most desirable guest.

Gonzalo talked to Wilson about Cuba's relationship with the United States. He summarized his views and those of many fellow Cubans as follows: "We deeply appreciate all that the United States has done for us. We like you Americans. But we value our sovereignty above all else in the world, and we want nothing done to interfere with it, curtail it, or restrict it."

When Wilson returned to Washington, he was persuaded by David Fairchild to drop his first name; having been christened Frederick Wilson Popenoe, in 1916 he became Wilson Popenoe. Fairchild himself had previously dropped his own middle name. Wilson's brother, Paul Bowman Popenoe, dropped his middle name at the same time. Apparently they were all influenced by President Woodrow Wilson, who had been christened Thomas Woodrow Wilson, but dropped the name Thomas.

Chapter IV

Guatemala, 1916–1917: The Avocado

Horticulturally speaking, the avocado was Wilson Popenoe's first love. When Carl Schmidt went to Mexico in 1911 to look for avocados, Wilson wanted to go along, but since he was one of the few horticulturists who knew how to bud-graft avocados, he had to stay in California to propagate the scion material sent back by Schmidt to Fred Popenoe's nursery in Altadena. After Wilson joined the U.S.D.A. he wanted to visit tropical America to look for avocados. Both in California and Florida, there was a keen demand for new sorts that might extend the season in which avocados could be sent to the market. But the navel orange expedition to Brazil, then the mango project in Florida kept him busy until the summer of 1916.

He then began to promote the idea of a trip to Guatemala for the U.S.D.A. Although there was a lot of information available about Mexican avocados, little was known about the Guatemalan fruit. Wilson pointed out to his superiors that an extended field trip to Guatemala to examine seedling avocados in dooryards and gardens in that country should provide some valuable new varieties for propagation in the United States. His project was approved, and in mid-September he boarded a United Fruit ship bound for Puerto Barrios on the Caribbean coast of Guatemala.

Puerto Barrios was a hot, insalubrious, malodorous banana port which was still growing. The sprawling town consisted for the most part of a cluster of ramshackle houses on stilts with *lámina* (corrugated iron) roofs and some shabby native shops, cantinas, and numerous shanties. The dingy waterfront was lined with coconut palms, the streets were mucky and filthy. Ominous black vultures called *zopilotes* were hovering overhead. It was like entering a country through the back door, for this unattractive port gave no indication of the charm of the rest of the country, especially the cool, distant Guatemalan highlands—the land of eternal spring.

Wilson spent the night in the best place in town—the clean, comfortable quarters of the United Fruit Company. Early the next morn-

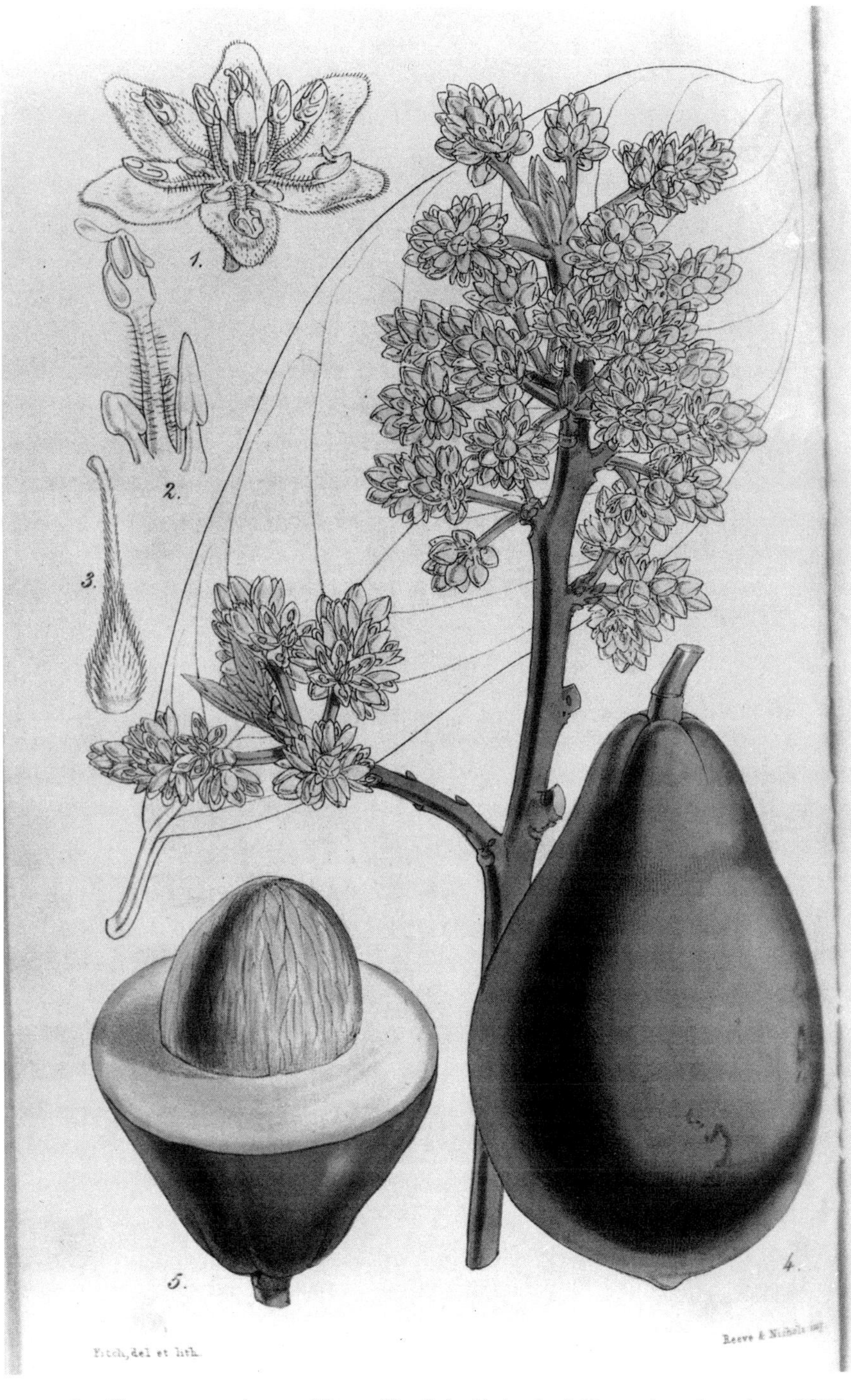

The avocado, *Persea americana.* (*From* Curtis's Botanical Magazine, *London, 1851. Courtesy of Botany Libraries, Harvard University, Cambridge, Mass.*)

ing, he took the narrow-gauge railroad for the scenic eleven-hour trip up to Guatemala City. Chugging along at twenty miles an hour, the train plunged first into a lush, moist tropical forest, featuring giant ceiba (silk cotton) trees, sacred to the Mayas, which towered over the tangled jungle vegetation. After an hour or so it emerged into the sunlight and the immense, well-kept United Fruit banana plantations. Here and there substantial wooden dwellings of laborers and overseers were to be seen. Gringos—banana men in riding breeches and high boots—were conspicuous around the stations where the train stopped every few minutes. During these stops, Indian women would appear balancing heavy baskets on their heads laden with local produce, including black beans, tortillas, tamales, cashew nuts, sweets, bananas, pineapples, and other fresh fruits. With shrill cries, they promoted their wares.

The train, following the banks of the muddy Río Motagua, passed Quiriguá where Wilson strained his eyes in vain for a glimpse of the famous Mayan ruins. He noticed the prominent United Fruit hospital crowning a hilltop, where he himself was to receive beneficial medical attention some years later. Another hour and the bananas were left behind. The country opened up as the tropical forest was superseded by scrub growth, thorn bush, cacti, and dust. The sun beat down mercilessly—the Guatemalan ladies sighed as they fanned themselves. At Zacapa, the train stopped while the passengers refreshed themselves and had some lunch at a local hotel. Two hours more across a parched and rainless plain, and the train left the river to begin the long climb up to the capital. Scrub oaks and pine trees appeared, while the temperature dropped noticeably. Corn fields were to be seen even on the steepest hillsides, some of which were at least forty-five degrees. Along nearby rushing mountain streams, cane and bananas occupied the vegas. It was nearing dusk when the groaning of the brakes and decreased speed of the train gave notice it had finally passed over a very long bridge, the *puente de las vacas,* and was descending onto the plain of Guatemala City, five thousand feet above sea level. The distant cones of the volcanoes Agua and Fuego were outlined against the sunset, an unforgettable sight.

Next morning, after a refreshing sleep in the cool air of the highlands, Wilson awoke to find an Indian woman at his door urging him to buy a clump of handsome orchids for fifteen cents. He then realized how much he liked Guatemala whether he found any avocados or not. He went out and strolled along the streets, so pleased with the climate and the friendly people he did not notice the roughness of the cobblestones or the fact they were missing altogether in many places. He was impressed by the mingling of barefoot Indians in their colorful costumes with people of Spanish blood in modern dress. He stayed for a few days at the home of Pedro Bruni, the United Fruit Company agent in Guatemala City.

It was the season of heavy rains, and away from the center of the city everything was deep in mud. The stage coaches or *diligencias,* which went daily to the old capital of Antigua, twenty-six miles away, did not always reach their destination. So Wilson visited Antigua by horse, a trip of some five hours. (Today the same trip takes about thirty-five minutes by auto.)

During the next sixteen months, Wilson rode over three thousand miles on horseback throughout Guatemala, searching for avocados. In Antigua, a region famous for its avocados, he found it was too early in the season for a study of the crop, since the trees, which were so prevalent in dooryards and coffee plantations, were bearing fruit not yet half grown. So he decided to visit the Verapaz, in the central part of the country, where the avocado was also renowned. In those days, there were two ways of reaching Cobán, the capital of the remote and verdant region known as the Alta Verapaz: one was to go overland from Guatemala City, a ride of about three and one-half days by horse; the other was to go back to Puerto Barrios by train, across to Livingston by launch, up the Río Dulce and across Lake Izabal on another launch, then by train from Panzós to Pancajché, a point not far from Cobán. Wilson decided on the latter route. He had several letters of introduction including one from Carl Vrooman, acting secretary of agriculture of the United States, which began as follows:

> To all who shall see these presents greeting: Be it known that Mr. Wilson Popenoe, Agricultural Explorer, in the Office of Foreign Seed and Plant Introduction, Bureau of Plant Industry, of the United States Department of Agriculture, a Citizen of the State of California and of the United States of America, has been authorized by me to make a study of the wild and cultivated plants of the Republics of Guatemala, Honduras, Nicaragua, El Salvador, and Costa Rica, for the purpose of ascertaining in what way a mutually beneficial exchange of seeds and plants may be brought about between those Republics and this Country.

Several days later, he reached Panzós, a tiny hamlet in a hot, malarial region, on the banks of the Río Polochic. It was raining. Mr. Guthrie, the manager of the Verapaz railroad, led Wilson through the mud to his house, half a mile from the river. They sat down together and ate a tin of cold corned beef for supper. Guthrie ran one of the

Happy New Year from Guatemala, January 1917. (*Courtesy of American Heritage Center, University of Wyoming.*)

most unusual railroads in the world. Twenty-eight miles long, it seemingly started nowhere and ended nowhere. In point of fact, it served to bridge a gap in the difficult transportation of coffee from Cobán, in the Alta Verapaz, to the sea. Freight was its major business, and Guthrie didn't worry about the passengers, no matter how much they complained. The weekly train started when he got ready, and came back when he got around to it.

After a night's sleep in Panzós, Wilson started out alone on muleback for the coffee *finca,* or plantation, of Kensett Champney, an American from Boston who had lived for many years in Guatemala and had carved out a little empire of his own in the wilderness of the Alta Verapaz. It was a beautiful ride through the Central American forest. The trail wound its way upward, crossing tiny but crystal clear streams. Here and there orchids were to be seen in the trees, including the striking *Lycaste skinneri,* whose albino form, the "monja blanca" or white nun, has become the national flower of Guatemala.

It was after dark when he reached the Champney residence. Several times he lost his way, but when he asked directions, he found out the Kekchi Indians didn't speak Spanish. He had to wait until a mule train came down the trail; the *arrieros,* or mule drivers, invariably spoke Spanish as well as the Indian tongue. Wilson was impressed by the cleanliness and attractive faces of the Kekchi women he met on the trail. They wore white *huipiles,* or blouses, worked with geometric patterns and colorful animals in blue, green, yellow, and red. Their skirts were all of the same white-checked indigo blue. (Every Indian village in Guatemala has its own typical dress or costume, like "the old school tie.")

Champney received Wilson cordially at Finca Sepacuité. A handsome man in his early seventies, with a flowing white beard, he spoke English with a Back Bay accent. He produced annually about eight thousand bags of coffee of 110 pounds each. His cost of production was five cents a pound, and he sold the coffee for from ten to twenty cents a pound, so he made a good profit—but he didn't care about dollars and cents since he literally had left civilization behind him. He was happy among his Indians, who looked upon him as a father. His wife was Kekchi and prepared a delicious dinner of hot tortillas, black bean soup, beef, and *huisquiles* (chayote squash), accompanied by a bottle of the finest French wine. In his notes, Wilson described Champney:

> Dinner finished, he led me upstairs, by the light of a hurricane lantern which he set upon a Louis XIV chair. The floors were covered with the choicest rugs, and he had two Victrolas, the most expensive models made. Still by the light of the lantern, he played me operatic airs for half an hour; then we both retired.
>
> At first it all seemed utterly incongruous to me—this man who had cut himself off from the world in which he was brought up, but nevertheless clung to the trappings of our own civilization while living almost like an Indian. And then, as the days went by and I learned to know Kensett Champney better, my ideas changed completely. Here was a man who

San Cristobal, Alta Verapaz, Guatemala, 1924. Wilson Popenoe is being served "batido," a ceremonial beverage for special guests, consisting of chocolate seasoned with black pepper. He had an extraordinary rapport with the Guatemalan Indians, in this case, the Kekchis. (*Courtesy of Hugh Popenoe.*)

> was big enough to realize that those Maya Indians had a civilization of their own, something different from ours, even better in some respects if not so good in others. He was taking the best of theirs, clinging at the same time to certain elements of his own background which gave him satisfaction.

Ten days later Wilson left Finca Sepacuité. He had seen many avocados and gathered much botanical information from Champney concerning plants in the Verapaz—which the latter had learned from the Indians. Taking Guthrie's weekly train, Wilson went down to a station where he was met by an Indian with a fine mule, which he rode to the next finca he was to visit called "Chejél," managed by another American, Bob Hempstead. Gathering numerous specimens for his plant press, Wilson arrived quite late at "Chejél," but was given a warm welcome by Hempstead who was to become a close friend. The latter, originally from Mobile, Alabama, had already lived in Guatemala for many years. His wife, Lulu, was born in Cobán of German-English parentage. Their home was an ideal place to see the coffee planter's life at its best. Wilson observed that if he could start life all over again, he would choose to be a coffee planter in the

Wilson Popenoe in northern Guatemala, 1917. Wilson spent sixteen months in remote areas looking for promising avocados—traveling more than three thousand miles on horseback. (*Courtesy of Hugh Popenoe.*)

Guatemalan highlands. Hempstead, a kind man, was an ideal *patrón* for his Indian workers. No Simon Legree, he treated them fairly and did not abuse them in any way.

Wilson and Hempstead got up early. Shortly after sunrise, two well-groomed mules were brought to the door. On the slippery trails of the Verapaz, where it is said "it rains thirteen months of the year," the sure-footed mule is superior to the horse. (Many years ago, the author knew an American from Kentucky named Burge, who brought a handsome, high-spirited thoroughbred down to Guatemala to ride about his coffee finca. One day, the nervous horse, frightened by a snake, plunged off a narrow path down a steep barranca—and the unfortunate gringo fell headlong to his death.)

It was the coffee-picking season, and every Indian the finca could muster was trudging toward the *cafetales* (coffee plantings) with his basket under his arm. Gangs were formed, and started to work under the supervision of *caporales* or foremen. (The usual *tarea* or day's work was to harvest a hundred pounds of coffee beans. Experienced men could sometimes harvest two or even three tareas; women usually managed to collect one tarea.)

Later they visited the *beneficio,* or processing plant, to see the coffee drying on the patio (after the red-skinned outer pulp had been removed by fermentation and the beans had been washed). At the end of the day, while enjoying a Scotch highball on the broad veranda of the Hempstead home, they talked of happenings in "the States" while gazing contentedly down the green valley of the Polochic to the shimmering patch of silver in the far distance which was Lake Izabal.

The mountains of the Verapaz were breathtakingly beautiful. Densely forested in parts, elsewhere cleared for coffee plantations or small farms of the Indians, the region seemed to Wilson to represent the tropics at its best. The climate was pleasant and healthful. Crystal clear water went tumbling down the mountainsides in numerous tiny rivulets to join the Polochic River. There were Indians everywhere—friendly, clean, and attractive in appearance.

Wilson asked a Spanish priest what he thought of the Kekchi Indians. The padre replied:

> They are good people, but many among them are given to drunkenness. An Indian will stay in his thatched hut back in the mountains for a month or more without a thought of *aguardiente* [inferior quality rum]. Then he will come into town bringing a pig or some other article for sale. With the proceeds therefrom he will walk into the first cantina he sees and proceed to get drunk. When he is capable of consuming no more liquor, he will be shoved unceremoniously out the door onto the street. Usually his wife, waiting for him, will sit patiently by his side for a few hours until he is able to stagger home. He then forgets about *aguardiente* for a month or so, until he comes to town with another pig.

Cobán itself was an interesting mixture of Indians and Europeans. There were, of course, numerous *ladinos* as well (as people of mixed Indian and Spanish blood were termed in Guatemala). Back in the 1860s, Enrique Dieseldorff, Bob Hempstead's father-in-law, had come over from Germany to the Cobán region and was soon followed by many other Germans—some of whom went into business, while most bought inexpensive land and planted coffee. These newly arrived coffee planters, or *finqueros,* frequently brought their wives with them and built comfortable houses, German in design. In the course of time, these hard-working German finqueros became the largest producers of Cobán coffee. Everyone around Cobán seemed to grow coffee; even the small *ranchos,* or one-room huts of the Indians, nestled among coffee bushes. Many handsome dwarf palms called *pacaya,* whose flower buds were eaten like asparagus, were also to be seen. Camellia shrubs, roses, dahlias, and other flowering plants adorned the environs of Cobán.

Wilson Popenoe on his favorite horse, Starlight, with his assistant José Cabnal, Guatemala City, 1917. Ready for the trail, Wilson and Cabnal looked like Don Quixote and Sancho Panza. (*Courtesy of Marion Popenoe Hatch.*)

Wilson's reconnaissance of the Verapaz with Bob Hempstead convinced him there was much more work to be done in that region. But as in Antigua, it was not yet the season for the avocado. Hempstead sold Wilson a handsome white horse, named Starlight, which he rode on the long journey overland back to Guatemala City, and on many subsequent trips.

Back in the city, Wilson realized he would have to travel everywhere by horse. There were no automobile roads in those days two miles beyond the city limits. He would also need an assistant, so he hired a reliable Kekchi Indian named José Cabnal, who knew how to handle animals and was familiar with the Guatemalan trails. A Presbyterian missionary warned Wilson: "Don't put shoes on that boy. No Indian can stand shoes. His morale goes all to pieces the moment he gets them on. You must not shoe a full-blooded Indian."

Wilson and José made many trips together during the following year—Wilson on Starlight, José bringing up the rear on a small mule. After two to four weeks in the field, they would return to Guatemala City so Wilson could prepare avocado seeds and budwood to ship to

The Nimlioh avocado of Guatemala, 1917. The Nimlioh is a large and productive variety, with some fruits weighing as much as three pounds. (*Photograph by Wilson Popenoe; courtesy of Hunt Institute for Botanical Documentation, Carnegie Mellon University, Pittsburgh, Pa.*)

Washington, write up his notes, and eat some good food. Out on the trail, the provisions were meager, as he carried little weight: a rubber poncho, a light blanket, a cotton hammock, a camera, a little clothing, and some canned butter and chocolates. Occasionally the going was hard, and both felt the strain. Once when they had broken camp while it was still dark, as was their custom, and were winding their way up a steep mountainside, as the first strands of dawn appeared in the eastern sky, Wilson turned in the saddle and remarked: "José, at this season of the year, the nights are long, aren't they?" To which the philosophical José answered quite frankly: "Yes, patrón, and so are the days!"

Four times they rode to the Verapaz, twice through the cool western highlands to Quetzaltenango, the second city of Guatemala. They spent several productive weeks in the region of Antigua. Wilson's trips on horseback through the Guatemalan highlands known as *tierra fria* were a joy for him. He considered this period the high spot in his career as an agricultural explorer. In his notes, he described these journeys:

A heavy load of eight thousand avocados near Guatemala City, 1917. (*Courtesy of Hunt Institute for Botanical Documentation, Carnegie Mellon University, Pittsburgh, Pa.*)

> Day after day we made camp by the side of the trail, usually under one of the thatched shelters which were provided for that purpose. Often we shared the rude hospitality of these spots with a group of Indians, or with a mule-train. While José unsaddled the animals and tethered them out to graze, I slung my hammock and prepared a simple meal. If we were in the neighborhood of an Indian village, José would forage for eggs, tortillas, and perchance a bottle of honey or some other delicacy.
>
> Tired from the day's journey—we usually made a *jornada* [day's ride] of seven or eight hours—I tumbled into my hammock as soon as dark had settled down upon us, murmuring the words of the immortal Sancho Panza, "God bless the man who first invented sleep." José, on the other hand, never seemed to sleep. At least I rarely caught him in the act. Every hour or two I would hear him striking a match, to look at the Ingersoll watch I had given him to see if it was time to get up. About three he would rise and go out to bring in the animals and feed them a few ears of corn, so that "they would be happy," he explained. By the light of a fat-pine splinter, we boiled a cup of coffee and perhaps a couple of eggs; this, and a toasted tortilla, and we were off.

They scoured the Indian villages, looking for avocado trees. They often heard the remark "Ahí viene el gringo aguacatero!" (there comes the American avocado man). When they found one in fruit that looked promising, they had to interview the owner, and inform him they wanted to examine his tree—for an honest purpose. Sometimes this

A Guatemalan Indian girl enjoying a tasty meal of tortillas and avocados. (*Photograph by Wilson Popenoe; courtesy of Hunt Institute for Botanical Documentation, Carnegie Mellon University, Pittsburgh, Pa.*)

explanation had to be lengthy, but invariably permission was granted. From time to time, Wilson would find good avocados, take notes, then go back and examine them again—for there were thousands of trees and he wanted only the best avocados, with fruits of exceptional quality, for introduction into the United States.

By late autumn of 1917, Wilson had finally selected twenty-three avocado varieties from several regions of Guatemala. He photographed the trees and cut identification numbers into their bark for future reference. He tasted and described the fruits in minute detail. He cut twigs from the selected trees which he wrapped in moist sphagnum moss; he then mailed this budwood to Washington in tin tubes or heavy, oiled paper. This scion material was grafted onto rootstock seedlings already growing in Washington. The grafted avocado plants were shipped subsequently to the U.S.D.A. Plant Introduction Garden at Chapman Field, Miami, where they were planted out in an abandoned field of the airport. Wilson also shipped back to Washington thousands of Guatemalan avocado seeds to be used as rootstocks.

It was time to go home, but he hated to leave Guatemala. As he rode out of the valley of Antigua on Starlight, he turned back for a last, wistful glimpse of the familiar volcanoes, Agua and Fuego. Several days later, laden with a batch of additional avocado budwood, he sailed for New York, and reached Washington in time for Christmas.

Soon thereafter the California Avocado Association petitioned the U.S.D.A. to send Wilson west to give a talk about his expedition to Guatemala. The meeting was held in Riverside. Although nearly six feet tall, Wilson was as lean as a ramrod after all those months in the saddle, and weighed only 120 pounds. After his speech, he asked if there were any questions. A wizened old man arose. "Mister Speaker," he drawled, "could I ask just one question? You said that you went all over Guatemala, hunting for the best avocados, and you sampled all the good ones to make sure they were up to standard. How many of them fruits do you think you et down there?" Wilson's records weren't very definite on that point, but he had to hazard a rough guess that he must have eaten a thousand or more. The old man looked him over up and down. "My God," he said, as he sat down, "and they claim that them avocados are fattening!"

A sympathetic boss is most helpful to a young agricultural explorer when the going gets rough. The Popenoes, father and son, both admired and liked David Fairchild for his concerned encouragement of Wilson. In December 1916, Fred wrote a warm letter of appreciation to Fairchild:

> Thank you for your kind words about Wilson. He is very anxious to make of this, his first independent exploration, an unqualified success, and may even overdo it in the way of personal effort. I have cautioned him strongly along this line. He wrote feelingly and with great appreciation of your recent letter to him, expressing approval of his work and offering some words of encouragement. He says when a fellow is far away from home and friends and in an alien land among strange people, living on poor food, and working hard, a letter like yours is worth $10,000 in gold. . . . He thinks there is no one like you, and in this we entirely agree with him.

Chapter V

Mexico and Washington, 1918–1919: *Manual of Tropical and Subtropical Fruits*

February 1918. World War I. Wilson was very patriotic and felt ill at ease not performing some kind of military duty. For a while he thought of offering his services to the medical corps of the U.S. army. Subsequently an arrangement was worked out whereby he obtained a leave of absence from the U.S.D.A. and went to Mexico, ostensibly to look for avocados and other plants as a botanist for the University of California. Beyond that, he did sub rosa intelligence work as a lieutenant in the U.S. navy, checking on the activity of German ships near the Mexican coast, especially in the states of Veracruz and Chiapas. When Wilson signed up for naval intelligence, the recruiting officer asked him how many languages he could speak. Wilson replied "Five," and was told, "That's enough. The last feller that came in said he could speak nine. We told him anyone who could speak that many languages wouldn't do anything but talk. We turned him down."

Ships were at a premium, but he managed to book passage on the SS *Breakwater,* a small banana ship plying its trade between New Orleans and Veracruz. The passage was very rough. As he lay in his bunk sucking on lemons and nibbling on soda biscuits, the door of his tiny stateroom was thrown open violently by an agitated seaman. "Man overboard," he shouted, "give me a knife so I can cut loose this life preserver." Sick as he was, Wilson jumped out of his berth and handed over an expensive French tool knife. They rushed to the deck. Fifty yards away, struggling in the turbulent sea, a man was shouting for help. Suddenly the cords were cut. With a mighty heave, the seaman threw the knife at the drowning man and handed Wilson the life preserver. Luckily, a few minutes later a boat was lowered and the victim was saved. Apparently he was a Mexican passenger who had tried to commit suicide but changed his mind when he hit the water.

The tropical port of Veracruz was the starting point for Cortés in

The original Fuerte avocado tree in Atlixco, Mexico. In 1911, Fred Popenoe imported from Mexico budwood of this exceptional avocado variety, which Wilson Popenoe grafted at the West India Gardens nursery in Altadena, California. For many years the Fuerte was the leading avocado variety in the United States. (*Courtesy of Hunt Institute for Botanical Documentation, Carnegie Mellon University, Pittsburgh, Pa.*)

A Mexican laborer in Puebla, Mexico, about 1920. The hook was used to harvest avocado fruits. (*Courtesy of Hunt Institute for Botanical Documentation, Carnegie Mellon University, Pittsburgh, Pa.*)

his march to Mexico City in 1519. Four centuries later, Wilson found considerable unrest and lawlessness among the local population and much discussion of revolution against the government of President Venustiano Carranza. As recently as 1914, United States troops had occupied Veracruz for seven months for security reasons. The air was charged again with disquieting political uncertainty.

For a few days Wilson discreetly checked on German ship movements (Mexico was neutral during World War I), then took a train for Atlixco in the state of Puebla. Train travel in Mexico in 1918 was precarious. From time to time, bandits would lift a rail, stop the train, and at gunpoint relieve passengers of everything they owned except their underwear. Luckily this never happened to Wilson, although he said he always wore his best suit of underwear before he boarded a Mexican train.

Atlixco, a center for avocado growing, was a picturesque town, situated on a plain six thousand feet above sea level at the foot of a hill, with the snowy summit of Popocatepetl towering in the background. Many other fruit trees were growing there in addition to avocados, and abundant vegetables and miscellaneous food crops were planted in small, irrigated gardens nearby.

In the fruit world, Atlixco was famous for being the home of the outstanding avocado variety called Fuerte—which had been introduced to the United States in 1911 thanks to the combined efforts of Fred Popenoe, Carl Schmidt, and Wilson Popenoe. Carl Schmidt collected budwood from the Fuerte mother tree which was growing in the backyard of a house in Atlixco and shipped it to the West India Gardens in Altadena, California. This famous tree was owned by Alexander LeBlanc, a naturalized Mexican citizen born in France, who recognized the exquisite taste of his own fruit and kept it for his family. The Fuerte tree first bore fruit about 1908, and lived another forty years. By 1938, the Fuerte variety accounted for over three-quarters of the avocado acreage in California. The California Avocado Association gave due credit to the original Fuerte mother tree in Atlixco by placing a copper plaque near its base, which reads in Spanish and English:

THIS TREE
HAS THROUGH ITS PROGENY PLAYED A MAJOR
ROLE IN THE DEVELOPMENT OF A NEW INDUSTRY
IN CALIFORNIA, U.S.A.
IN TESTIMONY OF OUR GRATITUDE AND
APPRECIATION THIS TABLET HAS BEEN PLACED
HERE BY THE
CALIFORNIA AVOCADO ASSOCIATION
1938

The avocado, *Persea americana,* is an evergreen tree of the laurel family. The natural range of the cultivated species extends from the cool central highlands of Mexico to the hot jungles of northwest Colombia. The Fuerte variety, according to Dr. Nigel J. H. Smith of the geography department of the University of Florida, is the result of a chance cross between Mexican and Guatemalan avocado trees. By the time the Spaniards arrived in the New World the avocado was being grown from Mexico as far south as Peru, and had already been domesticated for over two thousand years. In the Aztec or Nahuatl tongue, the avocado in Mexico was called *ahuacatl*; the Mayas called it *on*; in Peru, the name was *palta.* Our word avocado is an Anglicized corruption of the Mexican *ahuacatl.*

The fruit itself has a high oil content of up to 30 percent, as manifested by its buttery, smooth texture; also a high caloric content which, although not desirable for diet-conscious North Americans, is a valuable asset for many impoverished Latin Americans whose food intake is marginal. It is a relatively new crop: large-scale commercial production in orchards has only been carried out since 1911, following the introduction of the Fuerte variety. By 1988, over sixty thousand acres of avocado were planted in California, with a production of some 150,000 tons of the delicate fruit annually. Florida is the second-leading state with about ten thousand acres. Mexico, most of whose output is consumed locally, is now the largest avocado producer in the world. The fruit is also produced in Chile, Israel, South Africa, and New Zealand, among other countries. The European imports now amount to over 100,000 tons per year. In France, the avocado has replaced the pineapple as the number two exotic fruit after the banana.

In recent years, the high-yielding Hass has become the leading avocado clone, taking the place of the Fuerte—which is still popular. The Hass variety was found by accident in 1935 in California when a graft failed and the hardy rootstock kept on growing to produce a large crop of excellent fruit.

The avocado has become an expensive luxury in Japan, where it is said a single fruit can cost ten U.S. dollars. (The author remembers as recently as 1970 the very best Guatemalan avocados could be purchased near Santiago Atitlán for two cents apiece from local Indian producers.)

Wilson was familiar with the folklore concerning the avocado, especially in Guatemala. He was told how the Indians, to stimulate the growth of their hair, would break an avocado in half and rub the pulp over their scalps. Caged song birds, if fed on avocado pulp, were reputed to sing more beautifully than before, while laying hens would double their production. The aphrodisiac value of the fruit was famous and taken seriously in some quarters, as Wilson was to find out. In 1916, while staying at a modest, respectable pension in Antigua, he decided to try a diet of avocados alone to find out if it could be considered a complete ration. Could a man eat nothing but avocados and maintain his strength day after day? He made up his mind to find out for himself. When he told Doña Matilde, the owner of the boarding house (who had two young daughters), of his plan, she was against it—and said it would be bad for his health. He went ahead anyway, as he related:

> With science on my side, I thought I knew more about avocados than did Doña Matilde with her superstitions. Next morning I commenced the experiment. I weighed out eight ounces of avocado pulp, ate it, and went

Wilson Popenoe in Tapachula, Mexico, November 1918. (*Courtesy of Hunt Institute for Botanical Documentation, Carnegie Mellon University, Pittsburgh, Pa.*)

about my business. Toward ten o'clock I began to feel hollow and took another dose. By early afternoon, I began to wonder if it would seriously upset the experiment if I ate a few soda crackers. After all, I could figure out the number of calories they contained and discount the results. But with grim determination I stuck it out. The next morning I was feebly

> spooning my breakfast out of an avocado shell when there was a firm knock on the door. I opened it, and there stood Doña Matilde. "I am sorry to disturb you," she said; "in fact I am overwhelmed with embarrassment, but I have come to ask you to move out." This was a distinct shock.
>
> "What have I done?" I asked in astonishment. "Nothing, as yet," she replied. "I know you are a respectable young gentleman. I am sorry, very sorry."
>
> "Then why should you throw me out like this? Have I not paid my bills regularly? I have never come in late at night and kicked up a row, like some of your other boarders. I refuse to go unless you give me some adequate reason."
>
> Doña Matilde hesitated for a moment. Then, as delicately as possible, she placed the situation before me: "I have told you that avocados are not good for you—that is, if eaten to excess. You know that I have two innocent daughters in this house. It is not prudent; it is not safe. Either you will stop living on avocados, or you will leave my house this minute!"

Wilson terminated the experiment.

After he had obtained much information about avocado growing in Atlixco in 1918, Wilson took the train to Mexico City, a place that interested him because of its magnificent remains of Spanish colonial glory rather than its current political and social problems. He was intrigued by the horticultural history of the Aztec nation. He pointed out that we usually think of a warlike race as having little time and perhaps less instinct for the cultivation of flowers. But here was a people who, in the two or three centuries preceding the Conquest, developed a military prowess that has rarely been equaled, and at the same time built stately gardens, and filled their picture writings so full of symbols of flowers, fruits, and trees that we cannot doubt their passionate devotion to the soil and its products. Their elaborate botanical nomenclature can almost be said to have been the precursor of the Linnean system, for plants resembling each other in major characteristics often had a common generic name, with the addition of a prefix or suffix to indicate the specific kind.

Wilson was impressed by the artistic and cultural achievements of the early Mexicans, by the deep love of beauty in the hearts of the native population. This has persisted through centuries of turmoil, and has expressed itself notably in latter-day Mexican painting and music. Popular Mexican airs, sung with so much feeling in Mexico, were popular in thousands of thatched huts from the Rio Grande down to Chile. Wilson liked especially to hear two or three men with guitars put all their souls into "Cuatro Milpas," "La Borrachita," or "Jalisco Nunca Pierde."

Following some discussions with the U.S. naval attaché, he had to leave the pleasant life of the Mexican highlands and return to the lower elevations of the state of Veracruz to continue his botanical studies as well as his intelligence work. He visited Papantla, an attractive place in the heart of the Totonac country. The clean, handsome Totonacs resembled the Mayas in many respects; they took pride in their native costumes and customs. Horticulturally they were noted for the production of vanilla; they continued to cultivate that crop just as they did at the time of the Conquest. The vanilla plant, *Vanilla planifolia,* is a large, green-stemmed, climbing, perennial plant of the orchid family, native to the humid rain forests of southeastern Mexico, Central America, and the West Indies. Vanilla is the outstanding contribution of the Western Hemisphere to the spices of the world.

When the Spanish conquistadores were in Mexico in 1520, they observed that the emperor Montezuma was drinking many cups a day of *chocolatl,* a beverage consisting of powdered cocoa beans and ground corn, flavored with *tlilxochitl* (ground black vanilla pods) and honey. For many years after the Conquest, vanilla cultivation remained limited to the Veracruz region of Mexico until it spread to other countries, notably Réunion, the Seychelles, Indonesia, and especially the Malagasy Republic (Madagascar), now the world's largest producer.

Wilson was not familiar with the vanilla plant. One day while riding through some rather unattractive country of low bush, he asked his Totonac guide where the vanilla plantations were. "There is one," he said, "we have been passing through them for the past half hour." Wilson dismounted to get a closer look at some young saplings of various sorts growing in profusion nearby. Upon the trunks of these young trees vanilla vines were climbing, inconspicuously, their fleshy stems clothed with only an occasional thick, oblong leaf. One had to look hard to see them. The Indians gathered the pods before they were fully ripe, just as the color started to change from green to yellow. The pods were then taken to villages in the neighborhood for natural curing and fermentation—a process known as "sweating." During sweating the pods are spread in the sun daily for several hours until hot, then folded in blankets to sweat until the following morning. As this sweating process goes on, fermentation causes the formation of the peculiar and very fragrant white, crystalline compound now known as *vanillin.* Following ten to twenty days of sweating, the pods are dried in the sun for many weeks. In this traditional method five to six months may be required to complete the curing and drying.

Wilson and his guide headed for Gutierrez Zamora, one of the principal emporiums of the Mexican vanilla industry. The entire town had the delightful aroma of a candy shop in the States.

A few days later Wilson sailed back to Veracruz on the Gulf of Mexico, then proceeded overland by horseback on a long and tiresome trip to Oaxaca. He traveled on to Tapachula in the province of Soconusco, in the state of Chiapas near the Guatemalan border. This ancient and fertile Aztec region, with rich volcanic soils, was noted for its coffee and cocoa production. In former times, large quantities of chocolate for Montezuma came from Soconusco. The cocoa tree, *Theobroma cacao,* is native to the lowlands of tropical America. The beverage was unknown to Europeans until the arrival of Cortés in Mexico in 1519. The use of cocoa by other than native peoples is of recent origin compared with the consumption of tea and coffee. Centuries ago, chocolate was so valuable that the cocoa beans were used as money among the Aztec Indians.

In Soconusco, Wilson came upon a fruit that he believed was destined to become important in many tropical countries. This was the ilama, *Annona diversifolia,* of the custard apple family. The small tree, which thrives at sea level in the tropics, was cultivated extensively around Tapachula. The fruit is the size of a large orange; its white flesh, with the consistency of firm ice cream, has a delicate flavor suggestive of pineapple and strawberry.

Wilson was in Tapachula on November 11, 1918, the day World War I ended. There were no celebrations by the Allied colony—just relief the war was over. His naval intelligence work having been completed, he was free to go home. A few days later, he wrote to his parents from Tapachula:

> The rainy season is over down here, and the air is getting clearer. Yesterday afternoon, just at sundown, we had a glorious view down into Guatemala. Santa María [volcano] stood up as clear as day, and on down beyond it I could clearly discern the symmetrical cone of the Volcan de Atitlán, which I have often observed from the heights between Guatemala City and Antigua. Viewing these familiar peaks from this side of the frontier gives me an inexplicable longing to go back to Guatemala and take to the saddle again.

It was time for Wilson to terminate his Mexican trip and return to Washington. After paying the Mexican train conductor the customary bribe or *mordida* of nine pesos, he was able to begin his journey home.

In January 1919, while Wilson was returning from Mexico, David Fairchild wrote a promising letter to Wilson's father, Fred, from the U.S.D.A. headquarters in Washington:

Wilson Popenoe in Mexico, packing avocado budwood for shipment to the U.S.D.A. in Washington, 1918. (*Courtesy of Hunt Institute for Botanical Documentation, Carnegie Mellon University, Pittsburgh, Pa.*)

> I have felt all along that Wilson was thoroughly devoted to this project of bringing in new plants and had determined to make it his life's work, as I have done, regardless of what might come. I had opportunities early in the game to get out of it and financially would perhaps be better off today had I done so, but there are other things besides money, as you will appreciate, particularly in these days when values are being so tremendously shifted. I consider that Wilson is in line for substantial promotion and will get it, and if I have any power I shall see to it that he is taken care of financially. You can rest assured of that. His very interesting and worthy bulletin on avocados, which has been delayed so long, will soon appear, and it will be, I assure you, a credit to him. It will be a great delight to see him back here again.

Wilson spent most of 1919 working as a bureaucrat at the U.S.D.A. office in Washington. He devoted his spare time to writing. In April 1919, *The Avocado in Guatemala,* Bulletin No. 743, sixty-nine pages, illustrated, by Wilson Popenoe, Agricultural Explorer, Office of Foreign Seed and Plant Introduction, was published by the U.S.D.A. This was Wilson's first important publication; according to David Fairchild it established him as an authority on the avocado, and was the best-rounded piece of work the office had ever turned out.

Wilson was indeed an ardent admirer of the avocado, which he thought in time would become as popular in the United States as the orange. In an unpublished manuscript he wrote in 1924 for *National Geographic Magazine,* he described the importance of the fruit in Guatemala:

> The avocado is indigenous in Central America. When the Spaniards arrived upon the scene, they found this fruit from the Mexican highlands southward through Guatemala, Nicaragua, Honduras, and Costa Rica into Colombia, Ecuador, and Peru. Everywhere it was esteemed most highly, both for its delicious quality and for its food value, which ranks above that of the great majority of other fruits, including the banana. Two towns in Mexico took their name from the avocado, and there was a hieroglyph to represent the tree in the picture writing of the Aztecs.
>
> The importance of the avocado as a foodstuff among the Guatemalan Indians of today attests the prominent place it occupied in earlier times. Three or four corn cakes (*tortillas,* they are called), an avocado, and a cup of coffee—these are looked upon as the constituents of a good meal. And it is surprising how much physical labor the Indians can perform on a diet of this sort. Meat is a luxury, rarely to be indulged in; its place in the dietary is filled, to a certain extent, by the avocado. Containing as it does 15 to 25 percent of fat, and 2 to 5 percent of protein, this is a fruit of unusual character, and one which seems destined, in the not remote future, to play an important role in the life of northern peoples. More and more the tropics are being called upon to supply food for the over-populated regions of the temperate zone, and the avocado is capable of furnishing elements which are essential, and which in this fruit are put up by nature in a form so attractive as to make them irresistible.
>
> The Guatemalan aborigines have clung much more tenaciously to their ancient customs, including their food habits, then have the Indians of the United States. Perhaps this is because they had more to which they could cling. For the Central Americans far surpassed their brethren of northern climes in the development of the arts and sciences. The northern Indians were nomadic, living mainly by the chase: the principal tribes of Central and South America were agriculturists of no mean order, who lived in substantially built towns and villages, and who developed remarkable ability along several lines.

In 1914 Wilson had decided to join the U.S.D.A. instead of going back to college to complete his university education at Cornell. Five year later, despite only one year of college at Pomona, he succeeded in writing a most erudite book entitled *Manual of Tropical and Sub-*

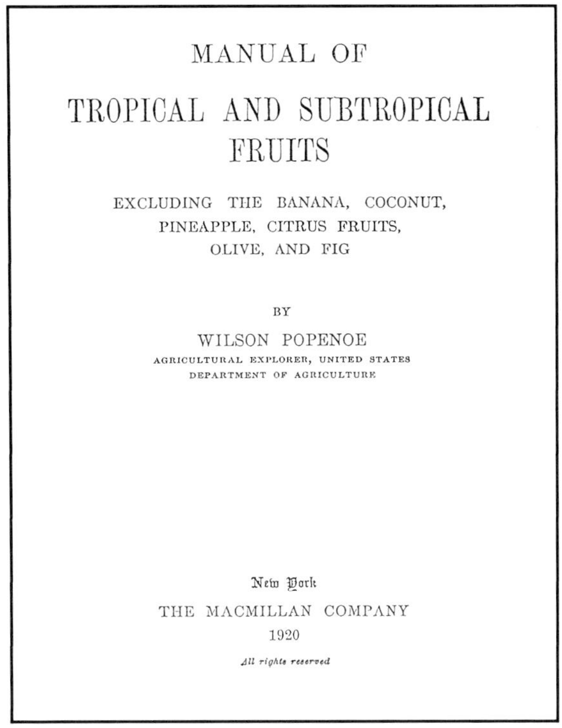
MANUAL OF

TROPICAL AND SUBTROPICAL FRUITS

EXCLUDING THE BANANA, COCONUT, PINEAPPLE, CITRUS FRUITS, OLIVE, AND FIG

BY

WILSON POPENOE

AGRICULTURAL EXPLORER, UNITED STATES DEPARTMENT OF AGRICULTURE

New York

THE MACMILLAN COMPANY

1920

All rights reserved

Wilson Popenoe had a brilliant mind. Although he completed only one year of college, this book, which he wrote in 1919, is still considered by many to be the standard work on tropical and subtropical fruits. (*Courtesy of Hugh Popenoe.*)

tropical Fruits: Excluding the Banana, Coconut, Pineapple, Citrus Fruits, Olive, and Fig. In this comprehensive work of 474 pages, he covered the principal tropical and subtropical fruits, dealing in detail with the avocado, the mango, the loquat, the guava, the litchi, the mangosteen, papaya, date, cherimoya, and many others. Wilson had traveled for seven years as an agricultural explorer for the U.S.D.A., and had taken advantage of the opportunity to gather at first hand information about these fruits. He pointed out in the preface that he had visited Hawaii, Japan, the Philippines, Singapore, Malacca, Penang, India, Arabia, North Africa, Mexico, Guatemala, the West Indies, and Brazil. He was also familiar with the cultural problems of tropical and subtropical fruit growing in California and Florida. He had been assisted and encouraged during the preparation of his manuscript by Charles Fuller Baker, under whom he had studied at Pomona; and especially by David Fairchild, to whom the United States was indebted for the introduction of many choice varieties of the mango, the date, and other tropical fruits.

First published in 1920 by the Macmillan Company, New York, the book subsequently had many printings, including a reprint by Hafner in 1974. Although somewhat outdated, it is still considered by many to be the standard work on the subject. Wilson stated in the original edition: "My intention in preparing the present work has been to bring together, for the guidance of those who live in the tropical and subtropical regions of the globe, the available information concerning the principal fruits cultivated, or which may be cultivated in those regions."

The book sold for five dollars. One reviewer observed in 1920: "Five dollars may seem quite an investment in a book, but when one sees it, considers its contents, and its authorship, the ghost of price quickly fades away. No one having an interest in the subject matter can afford to be without this peerless work."

In his introduction to the Hafner reprint edition of 1974, Wilson pointed out the main value of the book to present-day horticulturists lies in the fact that it is a unique source of background information. During the fifty years since the original publication, much progress has been made, but there is still a lot of work to be done:

> The horticultural varieties (cultivars they are called today) of the avocado and mango . . . have been superseded in nearly all cases by ones which produce better commercial characteristics, if not always better fruit. Indeed, readers should be reminded that the most popular commercial varieties of today may not be mentioned in a Manual of Tropical Fruits written in the year 2024. New varieties which originate as chance seedlings appear all the time, and the plant breeders are constantly seeking to produce new varieties capable of consistently bearing large crops, or resisting certain diseases, or extending the ripening season.

Wilson stressed the importance of vegetative propagation, usually some form of grafting—rather than the planting of seeds—since the seedlings of fruits improved by man through selection show too much variation in quality and character to make their commercial cultivation profitable.

Chapter VI

Costa Rica and Colombia, 1920

In early 1920 Wilson Popenoe departed from Washington on an extended plant exploration trip for the U.S.D.A. that was to take him to Guatemala, Costa Rica, Panama, Colombia, Ecuador, Peru, and Chile.

In Guatemala, his expedition got off to a bad start: while collecting plants in the Zacapa region, he cut his foot, which became infected. He reported to David Fairchild in March 1920:

> For this past month I have been breaking my neck trying to clean up my Guatemalan work so as to get down to Costa Rica by the end of March. But I fell into the hands of the United Fruit doctors at Quiriguá last Sunday, and you know what a thirst for blood those fellows have. They laid me out on the operating table, and washed their hands with alcohol and shot me full of cocaine, and then cut out a big chunk out of the sole of my right foot, and put four stitches in it, and then told me it didn't hurt.

Wilson had just returned from another of his horseback trips to the Verapaz with his usual complements of worn-out boots, muddy clothes, and broken saddle straps. He figured it took about a week for his clothes, his leather, and his body to recover from a journey to the Verapaz. David Fairchild had warned him previously on several occasions to conserve his strength or traveling over three thousand miles on horseback in the tropics would use up his young life. In May 1920 Fairchild cautioned him once more: "I am worried lest you go ahead too fast on your lame foot. Do not do it. Remember it is the only other foot you have, and you need it. Stop when you get to the right kind of a stopping place and have a competent physician pass on its condition."

But Wilson had to keep on the move. His foot gradually healed when he reached Costa Rica. He wrote another letter to Fairchild concerning the restrictions of time, which frustrated him as an agricultural explorer. Just when he was getting to know Guatemala well, for example, it was time to leave for Costa Rica, and soon thereafter for South America. He pointed out that to explore properly, one should visit a given place at different times of the year, so as to "spot" things when in flower and then return later to get seeds of them. But he had to travel constantly, or his trip would be endless.

In a way, it was a good time to leave Guatemala. A revolution was in progress that would lead to the overthrow of the dictator Manuel Estrada Cabrera, who had ruled the country with an iron hand since 1898. Wilson didn't worry much about revolutions in Latin America, however, as he pointed out in a letter to Fairchild sent through the diplomatic pouch:

> Things have looked a bit serious at times, but I haven't felt the least concern. Since being in Mexico, my favorite diet is revolutions. The Americans here in Guatemala City have had several scares, especially the dear ladies, who delight to make themselves believe they are all going to be killed, drawn, and quartered; but the fact is, the probability of any foreigners being hurt is extremely remote. There has been a little shooting among the participants in the affair, but not enough to make it really interesting. They caught me once in Antigua, on the day of the blowup over there, and seemed to think they had a prize; they went through me from head to foot, and finally struck what they considered was going to prove damning evidence in the form of a Unionist (the revolutionary party) propaganda sheet; but it turned out to be a religious tract that a missionary had just given me, and they finally had to let me go, rather to their chagrin.

In Costa Rica Wilson met a botanist and pharmacist named Otón Jiménez. They formed a lifelong friendship. Years later Jiménez described their first plant explorations in a memorandum he wrote in 1957 in Spanish, which may be translated as follows:

> In May 1920 Wilson Popenoe came to Costa Rica. We were both young fellows and both loved plants. On horseback and on foot we explored the steepest and most remote regions of Costa Rica for several months. We didn't mind the bad roads, the heavy rainstorms, the worst food imaginable, the mosquitos, not even sleeping on the floor of flea-infested huts. The satisfaction and pleasant surprise we experienced when we found either the plant material we were looking for or some other interesting new plants kept our spirits high. I was amazed how Wilson was always good-natured and cheerful no matter how uncomfortable and primitive our living conditions were.

On the lower slopes of the active volcano Irazú, near La Palma, they found several wild avocado trees, *Persea* sp., known as "aguacate de anís," since the gritty fruit tasted strongly of anise and licorice.

Wilson believed this was a primitive form of the Guatemalan race of avocados.

They collected seeds of and wrote an article about the pejibaye palm (*Bactris gasipaës*), a staple food crop of the ancient inhabitants of Costa Rica. The mealy flesh of the fruit, when boiled, tastes like sweet potatoes. They also collected fruits of "chayote" (*Sechium edule*), also known as "guisquil," a perennial vine whose fruits and young shoots were eaten as a vegetable in Central America.

Wilson found Costa Rica—where military despotism was unknown--more civilized and less violent than Guatemala. Yet he preferred Guatemala. He wrote Fairchild:

> Guatemala is a most peculiar place; it is at once the dirtiest and most fascinating country I have ever been in. Whenever I am there I think how good it will be to get out of it, and when I am out of it I think how much I would like to be there. . . . I remember only the pleasant features: those frosty mornings in the highlands, and the picturesque Indians carrying their huge packs across the mountain trails, and the perfectly fascinating life of the Indian villages. Ah! but Guatemala has something which Costa Rica lacks!

By mid-July Wilson was in Colón, Panama. He talked with the commanding officer at Fort Amador, Canal Zone; with the department quartermaster of the U.S. army and the commanding officer, Fort Sherman, Canal Zone. All were in favor of planting more ornamental trees, shrubs, and tropical fruit trees in various parts of the Canal Zone. Wilson was in total agreement with them and recommended to the U.S.D.A. that extensive tropical introductions be carried out at the military posts in the Canal Zone, where facilities and personnel were available to take good care of the plants. He recommended planting large numbers of seedling mangos of the Philippine race, which would make good shade trees, require very little care, and bear excellent fruit.

Wilson felt he had reached a sort of milepost in Panama: Central America was behind him, and South America in front. At the end of July he traveled by ship to Santa Marta, Colombia; then by launch to Barranquilla and up the Río Magdalena for twelve days, by river steamers; finally by train to Bogotá. After many days in the oppressive heat of the lowland tropics it was refreshing to see a portion of the great snow-capped Andean range in the distance, as the train ascended. When he caught the first whiff of highland air he felt new life enter his body. At nightfall, when they eventually reached the Colombian capital, it had become downright cold. A Chilean on the train advised

Wilson to put on his woolen underwear, not to bathe in cold water, and always to cover his throat well.

At the end of August, Wilson reported to Fairchild as follows:

> Bogotá as a city is much more imposing than I had expected. It is a small edition of Mexico City with many excellent buildings and a glorious setting, right at the foot of a huge hill topped by a monastery. There is no end of style here—gentlemen wearing cutaway coats and top hats at 8 in the morning, and all the young bloods carrying sticks. And the "cops" wear uniforms just like those of our own police force, and the troops, some of them wear spiked helmets. I saw one chap walking his post this morning in front of the barracks, and every time he turned around he brought his heels together with a click you could hear exactly a block and a half away. . . . The hotels are good, there is an abundance of fruits and vegetables; the streets are well paved, and there are electric cars. When the Bishop comes out as he did this morning when I was down on the plaza, they ring a bell to warn you, and you take off your coat and hat and get down on your knees, facing him. I daresay this is the most Catholic city in America. It is overrun with priests and infested with beggars, though the latter are not, perhaps, so abundant as in Mexico City. It is beastly cold at night, and of course there is no heat anywhere, so you put on your coat and shiver all evening. But when all is said and done, it is fascinating, and I like it. I am at last in the heart of the Andes!

Wilson write an article for *National Geographic Magazine* (published in February 1926) about his visit to Colombia in 1920, entitled "Round about Bogotá." He pointed out how elegant and pure the Spanish language was in Bogotá. For grammatical purity and rhetorical finish it was unexcelled in America. There was a delightful, quixotic love of flowery speech. He cited the following example:

> I was strolling one morning through a quiet part of the city when I was approached by a lad of ten or twelve years, who had obviously lost his way. He was just such a youth as would have stopped me in the United States with, "Say, Mister, how do I get to 224 Center Street?" But did the young Bogotano couch his inquiry in any such commonplace words? He did not. Removing his cap and making a sweeping bow, he inquired with that dignity which is possessed by every Latin, no matter what his estate:
>
> "Podría su personalidad indicarme donde está la calle Venezuela?" (Could Your Worthiness indicate to me the whereabouts of Venezuela Street?) Unfortunately I was as ignorant of the surroundings as himself; probably more so. I regretted my lamentable inability to furnish the required information, in the best Spanish at my command.
>
> "Pues, agradezco la fineza de su merced, y perdóname," he said in parting. (I thank Your Grace for the courtesy, nevertheless; and please pardon me!)

Nearing Facatativá, Colombia, September 1920. (*Courtesy of Hunt Institute for Botanical Documentation, Carnegie Mellon University, Pittsburgh, Pa.*)

Wilson pointed out in this article how different the fruits and vegetables were in Bogotá from the beans-corn-squash complex he had known so well in Mexico and Central America. Here there were other staple foodstuffs: the potato replaced Indian corn to a large extent; while other unfamiliar root crops were in evidence such as the *arracacha* (*Arracacia xanthorrhiza*), which tastes like parsnips; the *cubio* (*Tropaeolum tuberosum*), a small, white insipid tuber; and the *hibia,* a slender, pinkish white tuber which, when stewed with sugar, yields a product strikingly like green applesauce.

One of Wilson's missions in Colombia was to find the giant blackberry, also known as the Colombian berry, *Rubus macrocarpus,* one of the largest berries in the world. Single specimens sometimes measure over two inches in length and an inch and a half in thickness. He found a number of these shrubs growing wild in the cold Andean fog near Fusagasugá, an overnight trip from Bogotá by train and on foot. It was necessary for Wilson and his assistant, Marcos, to spend an uncomfortable night in a hut at a remote place called El Peñón. He described the experience:

> Stretched upon a bench, wrapped in my inadequate blanket, I attempted to pass the night with the minimum of suffering. It was one of the longest and hardest vigils I have ever experienced. The room was not boarded

> up securely, and the chill, damp wind off the mountain tops came through on all sides. My bones ached incessantly. I finally arose at dawn. We made out a breakfast of sorts and proceeded to business. In the excitement of being at last in the very region where the giant blackberry had first been discovered, my discomfort was soon forgotten. Marcos led the way, machete in hand. A drizzling rain was falling, and the boggy ground beneath our feet suggested that this must be the usual condition. Moisture, moisture everywhere; everything was dripping with it, and the plants looked as if they had never seen brilliant sunshine. Suddenly Marcos stopped and reached forth with his machete to cut a branch which had crossed our path. Looking up, I saw that it bore upon its outer end several red fruits the size of large walnuts—undoubtedly the giant blackberries for which I was searching. I grasped the branch with a shout.
>
> We dug roots, picked fruits, and took photographs as best we could, for it was drizzling steadily. . . . The plants were abundant, and their long, prickly canes scrambled to lengths of 15 feet or more. Their purple flowers, an inch across, were only less interesting than the gigantic berries, which I found to be flavored like loganberries, but to have larger and harder seeds. We packed up our spoils and . . . departed.

Every week Wilson sent seeds and cuttings to the U.S.D.A. in Washington via the diplomatic pouch. He had to economize in his mode of living since he was only allotted three thousand dollars for expenses for the entire year 1920. He pointed out to Fairchild that although it was necessary for him to live very frugally on such food as rice and plantains, while oil-company employees were living like princes, it was probably worthwhile since he got so much pleasure from coming into contact with new plants in his exploration that might enrich not only U.S. agriculturists but agriculture in general.

The climate of several coffee-growing regions in Colombia seemed ideal and reminded Wilson of parts of Guatemala, such as Antigua. He observed:

> Perhaps it is vain to think that an ideal climate exists in this world; but it has always seemed to me that an elevation of 4,000 to 6,000 feet in tropical America brings with it conditions as nearly perfect as one could wish. Indeed they are almost too much so; for day upon day of balmy sunshine, and night upon night of crisp, clean air, just right for sleeping under two blankets, make one begin to wish for a first-rate thunderstorm, a heat wave, or almost anything else to break the monotony.

In a tavern on a mountain trail near Facatativá, in the highlands of Colombia, Wilson was approached one day by a polite beggar, a tall youth in ragged garments, who, drawing himself up in stately dignity, bowed and addressed him:

Wilson Popenoe taking a day off from plant collecting for a picnic, or "día de campo," Cajicá, Colombia, 1924. (© *1926 National Geographic Society.*)

> Sir, I have need of your attention for a moment. You will, I know, pardon my boldness in thus addressing you, but I am going to Bogotá to have an operation on my leg and I am in need of funds. Far be it from me to ask for alms: but I have here some poetry—second-class poetry, it is true—which I am selling to defray the expenses of my journey. However, I am not going to ask you to buy it. I am going to make you a present of these verses, and you can tender me whatever the generosity of your nature suggests.

Wilson read the verses and although not impressed by their rhythmic quality was so touched by the ingratiating elegance of the indigent poet that he gave him ten cents.

The time had come to leave Bogotá. Wilson decided to cross the central cordillera of Colombia on horseback to the Cauca valley—a long, tiring journey overland. Looking for avocados, he found nothing of value—just a few inferior trees of the West Indian race. On the trail, however, he came upon a new species of papaya; large fruits with crimson-colored flesh and an applelike odor. The local inhabitants preserved them in heavy syrup to make the ubiquitous *dulce,* or sweet, so popular in tropical America. He thought this crimson-fruited papaya might be useful in Florida for breeding purposes. He also collected some yellow-fleshed potatoes of excellent quality, as well as several beautiful, heavy Cambria pineapples. Several days later he took a small steamer for a trip up the Cauca River. In a letter to Fairchild he described the countryside:

> As soon as I got on board I breathed a sigh of relief for she was as neat a little ship as you will find on Colombian waters, with real springs in the bunks and a shower bath on board. I began to feel better. And then the trip up the Cauca! Ah, but it was a joy! Never have I seen such fine river scenery. Really, the Cauca valley is a paradise, if there ever were on this earth. Groves of graceful bamboo along the river bank; vast pastures of Para grass in which the cattle stood so deep that only their horns and backs were visible; occasional picturesque little ranch houses along the bank; and now and then a cacao or coffee plantation with large trees for shade—some of them the beautiful *Erythrina umbrosa,* now in bloom. Yes Sir, those three days on board the river steamer were a constant joy: and how I slept of night! . . . I still think the spring bed is an improvement over the bare earth or barer boards.

At Cali Wilson terminated his Colombian work, dispatching a bag of seeds and plants in the diplomatic pouch. He then proceeded to the tropical port of Buenaventura, where he boarded a U.S. tramp steamer bound for Guayaquil, Ecuador, a trip that lasted three days. The boat was not allowed to carry passengers, so the captain signed Wilson on as assistant purser.

Chapter VII

Ecuador, Peru, and Chile, 1921

Shortly after his arrival in Ecuador, Wilson inspected an unusual collection of rare tropical plants on a neglected cacao hacienda called Payo, some twenty-five miles from Guayaquil. He found oil palm (*Elaeis*) in bearing, mangos from Jamaica, many ornamental palms, rambutans, litchi, carambolas with star-shaped fruits (*Averrhoa carambola*), and several Malaysian mangosteen trees (*Garcinia mangostana*), which were producing a good crop. But it was a familiar, sad story in Latin America. The wealthy Spanish owner, who diligently developed this fine collection, had died two years earlier. The children, having moved to Paris, displayed no interest whatsoever in the family estate their father had built up in Ecuador, and were trying to sell it. In the interim, uncared for, these exceptional palm and fruit trees were gradually dying out.

After a few days in the heat of Guayaquil "sudando la gota gruesa" (sweating the large drop), Wilson took the train up to cool Quito, elevation ninety-three hundred feet—a dusty trip of two days. He saw his first llamas and was delighted to note that Ecuador, unlike Colombia, had not Europeanized its Indian population. The Indians, dressed in picturesque native ponchos, were conversing in the Quichua tongue. At one point some Indian women swarmed alongside the train, selling roasted guinea pigs. But Wilson lost his appetite when he noticed the cold, stiff little bodies with wide-open mouths and two rows of teeth grinning at him. A few hours later he had recovered sufficiently to enjoy a generous portion of delicious Chilean strawberries.

He found Quito somewhat shabbier and more remote than cosmopolitan Bogotá, but the Ecuadorian people seemed more friendly. There was less emphasis here on getting dressed up on the city streets: he saw no Ecuadorian gentlemen in frock coats, silk hats, and carrying silver-headed walking sticks as he had in the Colombian capital. Ecuador was a down-to-earth, primitive country with a lot of old Spanish colonial atmosphere and he liked it. He enjoyed walking about the hilly streets of Quito, and admiring the exquisite churches. He just wished they wouldn't ring the church bells so constantly!

Wilson pointed out in an article in *Natural History* magazine in 1924 that the Quichua were skillful agriculturists who converted their mountainsides in Ecuador and Peru into productive land:

> On those rocky slopes they built series upon series of stone terraces, filled them with rich alluvium from the fertile valleys below, and irrigated them artificially from the mountain streams above. They brought many wild food plants into domestication and through conscious or unconscious selection carried some of them to a high degree of agricultural excellence, with the result that such plants as the potato and the sweet potato, the tomato, and the peanut are now cultivated and prized in many parts of the world.

Some of the hillsides were remarkably steep. He recalled one of his favorite stories from the Verapaz in Guatemala, about planting on perpendicular slopes: Ramón met Miguel on the road and said to him, after the customary salutations, "How is your brother Pedro?" "Ah," replied Miguel, "Pedro is in bad shape; he is so badly bruised that we fear he may not pull through." "What happened to him?" inquired Ramón. "Why, he was hoeing his cornfield, and fell off!"

Wilson made several trips by train to Ambato, the center of fruit growing in the highlands of Ecuador—a dry, cool region, subject to occasional frosts. He noticed a frost-resistant species of papaya (*Carica*), with a highly aromatic flesh. These papayas, known locally as *babaco,* were cooked to produce a dulce. He came upon some wild walnuts, *Juglans honorei,* handsome timber trees whose nuts had very thick shells. Most of the avocados in the Ambato region were of the Mexican race. He pointed out that the Spanish priests and friars, who came to Latin America with the conquistadores, established the best European fruits and food plants wherever they went, thereby performing a useful service. They also transported native species such as the avocado to regions where the plants had not previously been grown. Many capulín trees, *Prunus serotina,* were also growing around Ambato. This plant, which produces large, juicy, sweet black cherries, had also been introduced to Ecuador from Mexico by Spanish padres.

He observed the commercial plantings of the large-fruited, firm, flavorful Chilean strawberry, *Fragaria chiloensis,* which had been brought to Ecuador from Chile by the Spanish. It grew best in dry, sandy volcanic soil without irrigation. In 1714 a Frenchman named Frezier introduced a few plants of this Chilean berry to Marseilles.

Their progeny, when crossed with the imported, small-fruited North American strawberry, yielded for the first time the large-fruited strawberry commonly consumed in Europe and the United States.

Returning from Ambato on the train, Wilson enjoyed a splendid view of the snow-covered peaks of the Ecuadorian Andes. At one point the train passed near the base of towering Chimborazo; snow fields were just a few miles away. It was magnificent sightseeing except for the dust.

During March and April of 1921, Wilson made a lengthy and strenuous trip to the valley of the Río Malacatos in the southern part of Ecuador, near the Peruvian border. He wanted to observe the cherimoya, also known as the custard apple, *Annona cherimola,* in its native heath. It is said that centuries earlier Spanish padres had taken this fruit northward from southern Ecuador and Peru to Central America and Mexico, where Wilson had studied it. He had also seen it in southern California. With a smooth, white pulp combining the flavors of pineapple, strawberry, and banana, the cherimoya has been called a "masterpiece of nature," and ranks high among the best tropical fruits.

He traveled by train down to the port of Guayaquil, then boarded a river boat to the mining town of Santa Rosa. Proceeding on muleback with a mule driver, he passed through the Loja region where gold, silver, and copper were mined and *Cinchona* bark was stripped for quinine. This was rough, sparsely populated country. Travel by day was slow. It rained constantly, and the mules were often knee deep in mud. The nights were miserable: he slept in his hammock wherever he could find shelter—usually in some primitive hut where fleas swarmed upon him. Food was very scarce. One morning he opened his saddlebag to get a biscuit for breakfast, and found instead several hundred cockroaches.

After five days he came upon dozens of wild cherimoya trees in an indigenous condition on both sides of the trail, in the vicinity of Hacienda Chapamarca. He photographed many trees and collected seeds. This was an exhilarating treat for Wilson as an agricultural explorer and made the expedition worthwhile.

Wilson was near the canton of Paltas where it is believed that during the Inca invasion of Ecuador about 1450, the Inca ruler, Tupac Yupanqui, became acquainted with the avocado, which he carried back with him on his return to Peru. It is likely that the Quichua name *palta,* commonly given to the avocado in Peru, Bolivia, and

Chile (but rarely in Ecuador), owes its origin to this Paltas region in Ecuador, where Wilson expected to find many avocado trees, but saw few.

Continuing on the trail, Wilson reached some precipitous, mountainous terrain near a region called Oña, largely populated by Indians. He thought highly of the Quichua-speaking Indians, considering them to be the salvation of the Ecuadorian highlands. The country was too broken in many places to be farmed by machinery—too steep to use a plow, so that only the hand labor of the Indians could get anything out of the soil. He pointed out that those of mixed blood did not wish to get involved in that sort of hard work.

Several days later, Wilson reached Cuenca, altitude eighty-six hundred feet, called the "Athens of Ecuador," which had been founded by the Spanish in 1557, and where Panama hats were manufactured. He described his arrival there in a report to Fairchild:

> I tackled the last leg of the journey in excellent spirits. We were now in beautiful farming country. At 8 o'clock we were in the Tarqui section, which is one of the prettiest I have seen in all Ecuador—a small valley, the floor and slopes covered with corn fields, wheat fields and pasture lands. It was a regular spring morning with the birds singing, the sun shining and everything moist from recent rain. There were many people on the road, in striking contrast to the first part of this trip. About ten we came in sight of the ample valley of Cuenca, and just as the clocks were striking twelve we reached the plaza. Cuenca is the best town of interior Ecuador. By "best" I mean it is the cleanest, best-constructed, and in general the most livable of them all—Quito not excepted. The climate of Cuenca is famous—a "perpetual spring." And were it not for the fact that there is a long, hard ride over a mountain trail before one is in touch with the outside world (the railroad at Huigra) I am certain I would rather live in Cuenca than any other town in the Sierra.

Wilson had the good fortune to join a pack train of mules on his ride to Huigra—the final three days of his journey. They passed through fertile agricultural country, with splendid fields of corn, alfalfa, and barley. Most of the people were actively engaged in making Panama hats; he rarely passed a hut in which one or more persons were not sitting in the doorway weaving hats, including small children seven or eight years old. They rode upward to a damp forest region with unspeakably bad roads. It had been raining steadily for three months. The mules churned about in the mud, up to their bellies, but after several hours managed to get through. Suddenly on rounding a bend, Wilson saw Huigra lying in the valley below. He could scarcely refrain from shouting with joy. Huigra meant the railroad, the return to Quito, and freedom at last from mud, fleas, lice, wretched sleeping conditions, and poor, scanty food.

When Wilson got back to Quito, he wrote Fairchild a thoughtful memorandum containing some suggestions for gringos when they visit Latin America:

> You know in these Latin countries you've got to "conquistar simpatías" (win friends) and "hacerse querer" (make yourself liked) if you want to get along well. When a gringo comes here, refuses to learn the language, and criticizes everything Ecuadorian, he does not get very far. If, on the other hand, he shows a genuine interest in the language and institutions of the country, and expresses admiration for those things which he finds worthwhile, he is likely to find himself exceedingly popular within a very short time, and people, when they speak of him will say affectionately, "Ay que gringuito mas simpático!" than which there is no greater compliment.

He went on to make some special suggestions for agricultural explorers:

> The best policy, I think, is to enter a country quietly, with no press-agent's work ahead of you: call upon the Minister of Agriculture and those other functionaries whose aid you may need, in case of trouble; let them show you a few courtesies, in their own way, which will take a little of your time and will not achieve any practical results, but which is courteous and the right thing to do; and then pull out and go to work on your own hook. Lie low and dig in deep. Keep this up until you have finished what you came here to do, and then come out into the light once more. Make another round of calls upon the dignitaries, express your appreciation of the country's hospitality, its natural wealth, and their courtesies, and assure them of your desire to reciprocate in any way possible. Then come out in the local press with a few short articles, based upon some question of agricultural interest, in which you seize the opportunity to express your appreciation of the country's hospitality, natural advantages, and so forth. This puts you in right with the people at large. Then throw everything into your trunk, and jump on your steamer, and head for home.

At nine o'clock on a sunny May morning in 1921, Wilson started on another mule trip, this time to explore the Chota valley, some seventy miles north of Quito, and study the avocados of that region. He planned to obtain budwood from the most promising varieties to ship back to Washington in the diplomatic pouch. Accompanied by two mule drivers, who were whistling shrilly to keep their half-dozen pack animals from wandering into side alleys, he jogged down the main street of Quito on his mule, past the American Legation where he waved goodbye to the minister, the Honorable Charles S. Hartman.

At one in the afternoon they crossed the equatorial line, a narrow crevice in the hills nearby, which had been located and marked by a French commission several years earlier. Soon thereafter they reached the edge of the huge canyon of the Guaillabamba—a dry barren cleft in the earth's surface, three thousand feet deep. Some Ecuadorian friends had warned him: "Don't go near Guaillabamba, much less think of staying down there overnight. Why, people are known to have caught the most pernicious form of malaria just passing through that town on horseback." So they prudently slept near the upper rim of the canyon in a dilapidated hacienda house. Wilson had suffered from a bad case of malaria in Guatemala in 1917, and didn't want to tempt fate. In this remote place he met a young Belgian who had failed to present himself for military service during World War I, because he was afraid of getting killed. Now he was unable to get a passport to return to Europe. Living in exile, marooned as a "slacker," the Belgian wished he had gone to the war and taken his chances. He had not heard from his family for years, and lamented that his life was ruined.

Wilson awoke very early, at 2:00 A.M. The starlit heavens were clear, with the Southern Cross more brilliant than he had ever seen it. The mule drivers were already saddling their animals, muttering curses as they tugged on the pack ropes. He wondered if Latin American mule drivers ever slept; while on the trail he had often awakened in the night to hear a group of them, huddled together on a pile of saddle blankets, telling stories in low tones, laughing softly at a joke, while the red points of their cigarettes glowed in the darkness. Wilson and the mule drivers zigzagged down into the canyon, crossed the river, and began the long ascent to the village of Malchingui on the other side where they traversed the equator again. They continued to climb to the region known as the *páramo,* a cold, moist plateau at a high elevation where there was no other vegetation than grass or low brush. The road passed over the slopes of the volcano Imbabura, through tiny cornfields, barley fields, patches of potatoes, and other crops of the Indians whose primitive dwellings were scattered here and there on the broad slopes of the mountain. Wilson enjoyed sniffing once again the fragrant smoke of burning firewood filtering through the thatch of the Indian huts—a distinctive aroma he remembered well from the highlands of Guatemala. That night, they stayed in Ibarra in a wretched hotel, but Wilson was so tired after sixteen hours in the saddle he didn't mind the hard bed and the fleas. The following day, Ibarra was in a frenzy of excitement. The first airplane ever to

Wilson Popenoe on the equator, Malchingui, Ecuador, 1921. (*Courtesy of Hunt Institute for Botanical Documentation, Carnegie Mellon University, Pittsburgh, Pa.*)

reach that town was about to land on the newly constructed airfield, flying in from Quito.

Wilson went for a walk to photograph the marketplace, when he was abruptly arrested, taken to the police station, and relieved of his passport. It turned out that the chief of police was looking for a German thief who had stolen a lot of money and was believed to be hiding out in the vicinity. Wilson must have fitted the description of the Teutonic crook. Fortunately, he was able to get in touch with Minister Hartman, who arranged for his release through diplomatic channels. Wilson had the farsighted habit of calling on the American minister or ambassador as soon as he arrived in a foreign country, so Hartman knew him personally. The chief of police was disappointed: he thought he had made a big catch.

A few days later Wilson rode out of Ibarra with José Felix Tamayo, an Ecuadorian friend who had invited him to visit his splendid hacienda, La Rinconada, elevation 10,500 feet. It was long after dark and bitter cold when the horses' hoofs began to strike sparks from the cobblestone entrance to the hacienda. An excellent dinner awaited them featuring *locro,* a rich savory soup made of potatoes, egg yolks, milk, and most important of all, the native white cheese of Ecuador.

The next morning Wilson awoke to find himself in one of the most beautiful parts of South America. To the northward, the mountains rose to thirteen thousand feet, the higher slopes being of *páramo* character; while to the south the green pasture lands fell away toward the valley of the Chota, beyond which the snow-covered peaks of the Andes rose majestically. It was an extraordinarily fine hacienda. The owner's house was large and well-kept; off to one side there was a charming garden with attractive flower beds bordered by blue grass. José Felix had spared no expense in obtaining the finest seeds and flowers from the States. Sweet peas of the Spencer variety were featured. Wilson spent three days at La Rinconada, escorted by his quiet, warm-hearted generous host, looking for wild potatoes, wild blackberries, and flowering shrubs. One day they stopped for lunch at a mountaineer's hut, where Wilson was formally introduced to the guinea pig or *cuy,* as a tasty comestible—not as material for biological experimentation. He described this local pièce de résistance:

> In nearly every house throughout the higher regions of Ecuador you will find a considerable herd of these animals, their favorite pastime being to nose about your ears or run across your face at night when you are trying to sleep. They seem never to venture out doors; or if they do so, they do not get more than a few yards from their base of supplies, which is the

Wilson Popenoe with an Ecuadorian friend, José Felix Tamayo, near El Angel, Ecuador, 1921. (*Courtesy of Jorge M. Benítez.*)

zone immediate to the three stones which form the Andean cooking range. Potato parings seem to be their principal food, with cabbage leaves and other kinds of verdure occasionally thrown in for good measure. When nicely roasted they are excellent eating. I had always been given to understand, in the States, that they were not fit for human food, but that is

> all a mistake. They have a flavor midway between rabbit and spring chicken, and I am quite in accord with my good friend José Felix, who says that there is no better meat in this country than the guinea pig.

Wilson left La Rinconada with regret. It was an ideal Andean ranch, but he had to get on with his avocado survey. José Felix put Wilson in touch with Jorge Benítez, a local guide, who later worked for him faithfully for many years in Central America as an agricultural assistant, skilled horticulturist, and Jack-of-all-trades. For three days they wandered along dusty roads through the ancient haciendas of the Chota valley; they stopped at a run-down estate called San Vicente, formerly owned by Dominican priests, who, many years earlier had planted a fine orchard of avocados, including West Indian and Mexican varieties. Wilson found several promising trees and cut budwood to send back to Washington. Although the Chota valley had fertile, alluvial soil and an abundant supply of water for irrigation from the Río Chota, it was sparsely populated. The extremely hot, arid climate and the threat of pernicious malaria from the swarms of mosquitos from the swamps near the river made it feared by most Ecuadorians. Descendants of negro slaves brought from Africa by the Spaniards in colonial times made up the scant population, and kept the valley from utter abandonment. Sugar was the principal crop, made in the crudest possible manner and sold as coarse brown sugar in Ibarra. Wilson took precautions in these malaria-infested regions: he was equipped with a mosquito net and quinine bisulfate.

When they arrived at the town of Otavalo, where ponchos and carpets were manufactured, it was market day. Since it was late in the afternoon, the market was over and the Indians were departing in many directions. The men were dressed in brilliant red and yellow ponchos with white muslin trousers, the women in somber, dark blue garments enlivened by an abundance of silver bracelets, necklaces, and earrings. Many Indians were drunk, having consumed too much *chicha,* a fermented liquor made locally from corn or brown sugar, which had been their chief solace for many centuries.

Passing through Ibarra once again Wilson obtained seeds of sweet corn, potatoes, and native black walnuts for the U.S.D.A. In the local market he noticed there were several flours of different shades ranging from pure white to light brown: some were of wheat, some of barley, while others were of peas and other legumes. They passed through thinly populated highlands, with vast rolling pasture lands. When the Indians greeted Wilson, they said: "Alabado Sacramento" (Blessed

Plant exploration in Ecuador, 1921. The president of Ecuador conferred the decoration "Al Mérito" of the first class on Wilson Popenoe in 1921, in recognition of his services to Ecuadorian agriculture. (*Courtesy of Hunt Institute for Botanical Documentation, Carnegie Mellon University, Pittsburgh, Pa.*)

Sacrament), instead of the usual salutation "Buenos días, patrón." He was expected to reply "Por siempre alabado" (Forever blessed).

Returning to Quito after many days in the saddle, Wilson was delighted to brush off the dust, take off his puttees, hobnailed boots, and spurs, and settle down for a period of rest, relaxation, and report writing in more comfortable surroundings. Invited to a party at the American Legation, he put on his dinner jacket, called "un smoking." He wrote Fairchild that agricultural exploration in Latin America was in reality a Jekyll and Hyde existence. After roughing it in the back country for quite a while, under very primitive conditions, he thought he was undergoing a metamorphosis when he returned to the bright lights of the city, good clothes, excellent food served on a white tablecloth, a soft carpet under his feet, and shiny glass- and silverware around him.

The time had come to leave Ecuador: Wilson had completed his thorough avocado survey. In addition to preparing his detailed agricultural and travel reports, he had taken hundreds of photographs of avocados and other Ecuadorian fruits for the U.S.D.A. As a final courtesy he sent telegrams to many Ecuadorian friends and government officials, thanking them for all they had done for him—he was leaving the fine country with great affection for it, etc. ("Me despido cariñosamente de esta simpática tierra"). In early August 1921 he took a steamer from Guayaquil bound for Callao, Peru.

Having spent so much time in Ecuador, Wilson could only make a short visit to Peru. After a few days in Lima, where he paid his respects to several Peruvian agricultural officials, and, as usual, to United States diplomats, he traveled by train up to Cuzco, elevation eleven thousand feet. This ancient city, once capital of the vast Inca empire, has been called sometimes the "Archaeological Capital of South America," famous for its ruins of Inca fortresses, walls, temples, and palaces. The region is also noted for its ancient Inca agricultural terraces; Wilson visited the "staircase" farms of the Incas, which were still being used to plant corn, at Pisac and Yucay. These elaborate and marvelous terraces, including irrigation ditches, shored up the steep hillsides in immense flights of rough fieldstone terracing. Near Yucay, in the valley of the Río Urubamba, the Chilean strawberry was being grown on an extensive scale. Many strawberry and barley plantings had been made on the ancient terraces.

The most important Peruvian crop, however, was the white or Irish potato, *Solanum tuberosum,* which had been introduced to Europe by the Spaniards about 1580. Indigenous to the high Andes, where it has been cultivated since about 6,000 B.C., it was probably Peru's greatest gift to mankind. The modern potato is the fourth largest food crop of the world. It has been estimated that the world's annual potato harvest is worth many times the value of all the gold and silver of the Inca empire.

Other important agricultural commodities observed by Wilson in the Peruvian highlands included the root crop oca (*Oxalis tuberosa*), and the grain quinoa (*Chenopodium quinoa*). Both oca and quinoa, native to high elevations of the Andes, were consumed in great quantities by the ancient Incas and are vital staples for millions of Indians today. Both are little known outside their ancestral home, but are promising crops for the future in other parts of the world.

The colorful oca tubers, whose flavor is pleasant, resemble stubby, wrinkled carrots, with firm white flesh and shiny white to reddish skins. In the Peruvian highlands oca is the most important root crop after the potato. The bushy plant prospers in poor soils at high altitudes in the Andes.

The quinoa is an annual, broad-leafed herb which produces flat seeds, usually pale yellow in color; very high in protein, this grain was so vital to the Incas they considered it sacred. Highly nutritious quinoa is becoming more popular in the United States in health food stores and supermarkets. It has an unusually desirable balance of protein, starch, and fat.

It is interesting to note that in 1988, sixty-seven years after Wilson Popenoe had explored the Peruvian highlands, his son Hugh was chairman of a group of distinguished scientists who visited the same regions and carried on with the work Wilson had started. They produced an informative and worthwhile report entitled *Lost Crops of the Incas,* which describes in detail little-known plants of the Andes, such as oca and quinoa, with promise for worldwide cultivation in the future. The report was published by the National Academy Press in 1989, in cooperation with Noel D. Vietmeyer and the National Research Council.

Wilson arrived in Chile, which he called the "California of the South," on September 15, 1921. Spring was just breaking; all the deciduous fruits were in bloom. He planned to visit local nurseries and collect a number of small trees and plants for the U.S.D.A., representing the most interesting Chilean fruits. It would be autumn in the States, just the reverse of Chile, but his collection could be carried through the winter in greenhouses. The fruits he was interested in included avocados, apples, peaches, pears, grapes, oranges, grapefruits, cherries, olives, loquats, and Chilean strawberries.

Wilson's agricultural exploration trip to South America had lasted twenty months, and he was feeling tired. He wrote Fairchild:

> In spite of your insinuations that you doubt my ability ever to settle down, and all that sort of thing, I am much in hopes that I can, when I next reach Washington, take up more definite work there in the Office than I have heretofore had, and plan to limit my future trips to occasional explorations not more than 6 months in length. I do not believe that I ought to give up exploration: every once in a while there will be a definite piece of work to take up, such, for example, as a second and more thorough exploration of the Chota avocados in Ecuador, and if I can come out on some such trip from time to time I would like to do it. But never again, by all that is great and holy, do I tackle, alone, a two years exploration in which I attempt to cover pretty near all of the inaccessible, unlivable, and utterly impossible portions of the immense South American continent. No sir. I bit off a tremendous "chaw" this time, and while I think I am masticating it, my jaws are getting a bit fatigued from the exertion. . . . The peaceful comforts of a quiet life seem to me, just now, very attractive. Doubtless I will feel the Wanderlust stirring my blood ever

> and anon, but perhaps by spending a night on the ground somewhere in the Virginia woods, I can keep the thing in control. The safest cure I can imagine, would be to prepare a meal of undercooked black beans and clammy tortillas: one or two of these would, I feel sure fix me up at any time. I think the Duke of Wellington was correct, if, as reported, he said that "Habit is ten times Nature." I have never been able to see why I shouldn't be able to live comfortably on a diet of black beans, rice and tortillas, except on the grounds of Habit.

Wilson took the train from Santiago to Santa Inés to visit the large, commercial plant nursery of Salvador Izquierdo. After the dry and barren highlands of Peru, it was a refreshing change to gaze out the train window and see the green pastures, attractive gardens, cultivated fields, and well-kept, flowering fruit orchards of Chile. Many homes had potted plants in the windows, a rare sight in other parts of Latin America. Izquierdo, a leader in the Chilean fruit industry, had the finest horticultural nursery Wilson had seen since leaving the States. It was a joy for the plantsman in him to handle the small potted plants, run his hands through the loose soil along the nursery rows, and view the straight lines of little peach and pear trees in bloom. No doubt it reminded him of his father's West India Gardens in Altadena. Izquierdo also ran a canning factory which turned out first-rate canned vegetables and fruits, whose flavor, quality, and price compared favorably with California products. The factory put up asparagus, peas, string beans, and half a dozen kinds of fruits and jams.

He visited La Cruz, the most important avocado-growing region in Chile. At Quillota, near Valparaiso, there were irrigated peach orchards, with Chilean strawberries growing under the peach trees as a secondary culture, both crops in excellent condition. He called at the National School of Agriculture in Santiago, and was impressed by the strict discipline and thorough training of the students. He observed apple trees being sprayed with lime sulphur, and citrus trees planted experimentally with reed mats suspended above them for protection from the cold.

On October 12, Wilson sailed from Valparaiso on the Grace Line steamer *Santa Ana,* taking with him aboard ship, as planned, a large number of plants of selected Chilean fruits for Washington. The ship arrived at New York on October 31, 1921, as he completed his mission. Thus ended Wilson Popenoe's last extensive trip as an agricultural explorer.

Chapter VIII

Washington, 1922–1925: Marriage to Dorothy Hughes

David Fairchild used to ask Wilson from time to time when he planned to get married. It was a sort of standing joke between them. Just before leaving Chile in September 1921, Wilson wrote his chief:

> I do a lot of talking about Girls, don't I, for a fellow who isn't serious about the subject? I imagine if I had not been on the move so much of the time these past ten years, however, it would have been hard for me to keep out of deep water. As it is, I have had to do some fast running on two or three occasions, but my desire to carry out your wishes in the matter of early marriage has kept me from taking a serious interest in the subject. With me, it has so far been Girls rather than any particular Girl. One of these days I shall, most probably, focus my attention upon some particular one, but I don't intend to do so foolishly nor ill-advisedly. And when I get into the matter seriously I will inform you to that effect: in the meantime you can know that any wild rumors you hear about me are due simply to liking for company of the opposite sex, and for the dance. . . . You cannot imagine how keenly I am looking forward to seeing all of you again, and to settling down for the winter in Washington. It is going to be an interesting season there, I am sure, and I want to be on hand.

Wilson settled down in Washington following his lengthy agricultural exploration trip to South America. He was in good spirits. His annual salary at the U.S.D.A. had been raised to three thousand dollars. Furthermore, he was in demand as a lecturer. In January 1922 he gave a talk on Guatemala, with slides, in Brooklyn, New York, before an audience of about three hundred Brooklynites; at first he was nervous, but soon regained his confidence. For this effort he was paid a modest honorarium of twenty dollars. A few days later he lectured in Washington about his experiences in Guatemala at the National Geographic Society, to a much larger audience. This time, he was more at ease since he knew the subject so well. A witty raconteur, he told several jokes in his own inimitable way, which the crowd seemed to enjoy. Wilson received a standing ovation and a fee of two hundred dollars—the first significant payment he had ever received for lecturing. He enjoyed the lecture circuit, where his engaging personality, vivacious enthusiasm, and thorough knowledge of horticulture and Latin America seemed to captivate his audiences in Washington, New York, New England, and California.

David Fairchild, now fifty-four, had been an effective, highly regarded chief of the Office of Foreign Seed and Plant Introduction of the U.S.D.A. for many years. He himself had founded the office in 1898, and was responsible for many important plant introductions to the United States from all over the world. His wife, Marian, was a daughter of Alexander Graham Bell. Fairchild was thus enabled financially to spend his summers in Nova Scotia at Dr. Bell's summer home and his winters in Coconut Grove, Florida. When Fairchild was away from Washington--which happened frequently—he appointed an acting head of the office to represent him. Wilson Popenoe filled this deputy position intermittently from 1922 to 1925. It was an awkward role: although he was responsible for the details of all routine business at the office, he could not make any important decisions without Fairchild's approval. Many times, a decision had to be made, and Fairchild was incommunicado—away on a trip. During Fairchild's absences, there was frequent bickering among the office staff.

Wilson had only been acting chief for a few weeks when he began to realize he wasn't cut out for bureaucratic work. He wrote Fairchild in 1922:

> It is pretty hard to turn out papers when you have as much routine on hand as I now have, and I can begin to appreciate the difficulties under which you have worked for many years. My previous stays here have been altogether different in this respect. I can see that all the writing I do in the future will have to be carried out under rather difficult circumstances—mainly on Sundays and holidays—or by stealing a morning now and then, and crawling into some quiet hole where nobody can find me.

Looking on the bright side, Fairchild replied in a letter from Coconut Grove, Florida, dated April 10, 1922: "You don't know, dear Wilson, with what a feeling of security I look forward into the future knowing that you have put those energetic shoulders and that nervous energy and your winning personality into the net of red tape and the harness of fascinating things to be done there in the office."

Introduced around Washington by Fairchild, Wilson became a member of the elite Cosmos Club, where he would often dine after a trying day at the office. According to Knowles Ryerson, Wilson brightened their lives at the Cosmos Club, as he took a keen interest in the activities of his friends and associates. In the club, there was a small, unorganized group of active explorers who called themselves the "Outlanders." They gathered for dinner whenever a member was leaving on an expedition to give him a proper send-off. Wilson was the traveling foreign plant explorer member of the group.

Wilson enjoyed a very close and affectionate relationship with Dr. and Mrs. Fairchild, who looked upon him as a godson. He was often a guest at their attractive home, "In the Woods," in Chevy Chase, Maryland. Wilson also attended many soirees at the home of Alexander Graham Bell, which was a sort of cultural center. On Thursday evenings it was a social gathering place for intellectuals and high government officials—a counterpart of the traditional European salon. Wilson thus got to know many important people in the nation's capital, both in and out of the government—the cream of the scientific and cultural society.

During this period Wilson used to confide in Mrs. Fairchild about his love affairs, and ask her for advice. In July 1922, he wrote his godmother:

> You see, I met a very interesting girl, but she is, up to the present time, determined upon a career as a concert pianist. She has talent, I think, but not genius, and will never be famous. She was born and reared in Smyrna, where her father is a professor in an American college. She is of American stock, graduated last year from Oberlin, has gray eyes, dark brown hair and a fair complexion. Incidentally, she has the brightest and most penetrating intellect I have ever seen in a girl of her years. But she has this *career* in her head. Now come on with your advice, please!

Apparently Mrs. Fairchild was not too keen about "gray eyes." In any event, Wilson came to the conclusion the girl was too selfish and artistic in temperament for him, and thus the affair ended.

Then he disclosed to his godmother that there was another attractive girl named Helen, with blond hair, blue eyes, and a poodle. Wilson had bought a used Chandler Light Six roadster and used to drive around Washington and Baltimore with her on weekends. The trouble was, everywhere Helen went, she insisted on taking the dog along. Wilson didn't like the confounded poodle, so that was the end of the brief romance with Helen.

In July 1923, Dorothy K. Hughes, an attractive, intelligent young British woman, was hired by the Bureau of Plant Industry. Born at Ashford, Middlesex, England, in June 1899, Dorothy had studied at the University of London and worked for several years on African grasses in the herbarium at Kew. She had become a capable botanist

Dorothy and Wilson Popenoe on their honeymoon, Savannah, Georgia, December 1923. (*Courtesy of Nancy Popenoe Guillou.*)

and botanical illustrator. Dorothy and Wilson met at the U.S.D.A. offices in Washington. It was love at first sight. A few days later she called him and asked if he had any work for her. Wilson thought to himself, "Yes, I have a lifetime of work for you." He soon proposed to Dorothy and they were married in November 1923 at the Fairchild's home, "In the Woods." Over the next nine years, Wilson and

Wilson Popenoe and David Fairchild in Florida about 1932. David Fairchild (1869–1954) was a noted horticulturist, plant explorer, and author. Founder in 1897 of the U.S.D.A. Office of Plant Introduction, he was its chief for many years. He traveled throughout the world and personally introduced thousands of different species or varieties of plants to the United States. The Fairchild Tropical Garden in Miami is named for him. He wrote several books, including *Exploring for Plants* (1930); *The World Was My Garden* (1938); and *Garden Islands of the Great East* (1943). The palm in the photograph is *Chamaedorea pacaya.* Its unopened flower clusters, dipped in egg and fried, are eaten as a vegetable in Central America. (*Courtesy of Fairchild Tropical Garden, Miami.*)

Dorothy were to have five children: Peter (1924), Nancy (1926), Hugh (1929), Marion (1930), and Pauline, also known as Sally (1932).

Wilson and Dorothy settled down to married life at 1940 Biltmore Street in Washington. He was extremely busy at the U.S.D.A., where, in spite of the hectic atmosphere, a great deal of important work was being done. The collection and introduction of seeds and plants into the United States had come a long way since 1839 when the first appropriation of one thousand dollars for agriculture was authorized by the Congress. David Fairchild had been responsible for several outstanding plant introductions, including selected cotton varieties from Egypt, which after breeding and further selection contributed to the development of the cotton industry in the southwestern United States. He also played an active part in the introduction of hairy Peruvian alfalfa, an important forage crop; several species of clover, valuable in crop rotation; and widely cultivated sorghum grasses, used for stock feed and forage. All in all, under Fairchild's direction, more than sixty thousand lots of plants were accessioned—including field crops, fruit crops, and ornamentals, introduced from every quarter of the globe. Some, as noted above, were useful from the economic standpoint; other ornamentals surpassed in beauty and delicate coloring anything seen before in the United States. With farsighted vision, Fairchild organized future plant introductions of the U.S.D.A. in his bulletin No. 21 of 1898 entitled *Systematic Plant Introduction—Its Purposes and Methods,* which for many years provided guidelines to the Bureau of Plant Industry.

In a letter to Wilson Popenoe from England, dated October 19, 1924, Fairchild pointed out some differences between the U.S.D.A. setup in Washington and the Royal Botanic Gardens at Kew. These British gardens, located in a suburb of London on the right bank of the Thames and considered by many to be the finest in the world, originated in 1759:

> The old house at Kew has stood where it now stands for over one hundred years, whereas everything botanical you can put your hand on in Washington except the so-called Botanic Gardens has been moved about like a peddler's wagon. I told them here that my affairs and effects have been moved in toto eight times in 25 years and I think it rather amazed them. The herbarium at Kew with its wood floors, staircases, alcoves and stools maintains almost the quiet of a monastery. The quiet voices of the workers are in striking contrast with the high-pitched, strident notes of our American workers, and the total absence of the eternal click of the typewriter gives to the quiet of the place a perfectly heavenly character. It has a dignity compared with our helter skelter crowded offices and hurrying

noisy bustle. . . . Of course there is one thing which has contributed more than any other to our hubbub: it is our universal use of the telephone. There is one telephone in the herbarium but it is where nobody can hear it ring. . . . It makes me feel ashamed that I have not done more to bring about better working conditions in Washington.

Fairchild was overly modest about his achievements. Both he and the Office of Foreign Seed and Plant Introduction had much to be proud of, even though their modus operandi was not as tranquil as the British.

One day a Washington newspaper reporter came around to Wilson's office to interview him. After some routine questions, the reporter asked how an agricultural explorer goes about finding plants in the hinterlands of an undeveloped foreign country. Wilson replied:

You go to some forgotten town and hire a native boy. Then you buy three animals—horses, mules or maybe camels—one for yourself, one for the boy and one for the baggage. Then you head for the back country and keep going until you reach a place where there are so few cooking vessels that you can pay for a night's lodging with an empty tomato can. Then you go to the village market place and watch like a hawk for everything brought in to be sold. You almost make love to the natives. You get invited to dinner. And finally you get the plant or the seeds of the plant you want. If it is a plant, you pack it in moss. If there are seeds, you pack them in charcoal. You put the package on the pack-mule, give the boy $4 in American money—which often as not is about $200 in the money of his own country—and start him back. He represents himself as a great dignitary of the American government; and invests the $200 native money as his fancy dictates. One boy bought twelve pairs of earrings for twelve native girls. But the point is he gets your package off to Washington.

The reporter then asked Popenoe what was the most uncomfortable place he ever slept in. Wilson answered him as follows:

It was in a little village in Guatemala. I got lodging in a hut containing six dogs and three Indians. I swung my hammock between two rafters and went to bed. Before I could get out, the flea bites I had accumulated, if placed end to end, would reach from Washington to Baltimore. I counted eighty-five on one arm below the elbow and seventy-four on the other. But I got the plant I went after. It's here now, growing in the United States.

In a more serious vein, Wilson pointed out in the *American Foreign Service Journal* exactly how his department functioned. There were two means of securing plant introductions employed by the Office of Foreign Seed and Plant Introduction in the Bureau of Plant Industry: (1) through agricultural explorers sent out on special missions to obtain

Wilson Popenoe in his office at the U.S.D.A. in Washington, 1923. As acting head of the Office of Seed and Plant Introduction, he did not enjoy the life of a bureaucrat, sitting at a desk all day long. (*Courtesy of Hunt Institute for Botanical Documentation, Carnegie Mellon University, Pittsburgh, Pa.*)

certain plants or to collect promising plant material of all sorts in remote and little-known regions; and (2) through correspondence with diplomatic and consular offices abroad, as well as with experiment stations of foreign governments, botanists, missionaries, and others.

Plants introduced from these sources, once received in Washington, were submitted to rigid inspection. If free from dangerous insect pests and plant diseases, they were sent to one of the plant introduction gardens for preliminary testing and propagation. Careful records were kept. As soon as sufficient stock became available, young plants were sent to experiment stations of the U.S.D.A., and the various states, as well as to park superintendents, nurserymen, and others qualified to give an intelligent trial to the new arrivals.

Wilson did not care for the confinement, red tape, and bureaucratic atmosphere of the Washington office. From time to time he wrote Fairchild he was sick of clerical work and didn't want to get "clericalized." The bureau, however, was stressing office efficiency and administrative routine, rather than agricultural exploration. Wilson's expertise in foreign lands was belittled. He was told in no uncertain terms by two high-ranking bureaucrats of the U.S.D.A. named Kellerman and Allanson to run his office more efficiently and also that he seemed to be temperamentally unsuited for administrative leadership. Furthermore, he was warned that several of Fairchild's directives were not to be heeded, since they were no longer in force. This was the last straw. Basically a field man, not a bureaucrat, Wilson now felt like a fish out of water. On May 6, 1925, he wrote Fairchild as follows:

> This is a letter I had hoped never to write you. Things in the Bureau have reached the point where it seems useless for me to hang on any longer. I was trained under you, and I came into this work with the understanding that I was going to run it more or less along the lines you had followed. Now they don't approve of many of your policies. I have constantly struggled, these three years, to keep alive your ideals and your policies for the Office. It has been a losing fight. Rather than stultify myself by changing the whole course of affairs to meet the ideas of someone else, I prefer to step out. Then they can make of the Office the routine, colorless, cog in the Bureau wheel which seems to be their ambition, and I won't be here with a feeling that I have prostituted my ideals to "organization" and "efficiency."

Two months later, following twelve years of effective service as an agricultural explorer, Wilson Popenoe resigned from the U.S.D.A. to take a challenging and more lucrative job with the United Fruit Company in Honduras. His wife, Dorothy, was very much in favor of this move. In a letter to Mrs. David Fairchild, dated August 25, 1925, Wilson gave the following explanation:

> I have often realized that I was not cut out for office work in Washington. Had I stayed here longer I am sure I would have had to give it up eventually, and perhaps at a time when it would have hurt much worse than now. I am leaving for two reasons: first, I have not the physical and nervous strength to stand the daily round of office routine; and second, I have struggled hard to keep alive something of the Fairchild ideals and spirit, and these do not seem to find favor with certain of my superiors. I go with the feeling that I would rather have stayed here three years, and tried to carry on in the way Dr. Fairchild and I believe is right, than to have stayed here for life by giving up the things in which I believe, and doing the drab sort of work which is apparently going to be required.
>
> I went to Honduras, and found things quite attractive there. They have been very generous with me: they let me pick out the sort of house we wanted, and are now building it for us. They say I am to have a free hand, and will make my own job. I will of course go on with the sort of work I have been doing, the development of tropical fruits, and probably a few other new tropical crops. They are going to pay me $6,000 a year to begin with, in addition to furnishing our house, two servants, and all the heavy furniture. We get six weeks leave annually, with two passages to the states and back.

In January 1925, the Universidad Mayor de San Marcos in Lima, Peru, generally conceded to be the oldest university in the Western Hemisphere, conferred on Wilson Popenoe the degree of Doctor en Ciencias, *honoris causa.* This degree was significant: the University of San Marcos was founded in 1555, through a charter granted by the king of Spain, eighty-one years before the founding of Harvard University—the oldest in the United States—in 1636. After 1925, even though he had never completed his formal college education in the United States or anywhere else, Wilson was invariably referred to as "Dr. Popenoe" in Latin America, rather than "Sr. Popenoe" or "Mr. Popenoe." In the course of time, many people in the States also called him "Doctor."

Chapter IX

Lancetilla, Honduras, 1925–1940: With the United Fruit Company

During the last quarter of 1925, Wilson departed from Washington for an inspection trip to Honduras on behalf of the United Fruit Company. He selected the site for the new Lancetilla Agricultural Experiment Station, at the head of a small valley situated three miles from the Caribbean port of Tela on the north coast of Honduras. The project was to be controlled by the Tela Railroad Company, a subsidiary of United Fruit. The name of Lancetilla is derived from a small, spiny indigenous palm that grows throughout the valley, whose little spines look like lances.

In December 1925, Wilson brought Dorothy and little Peter down to their new home at Tela, which was on the edge of a beautiful tropical beach, lined with coconut palms. The main headquarters of the experiment station were located at Tela where substantial port facilities had been constructed, including the wharf, railroad yards, power plant, general office building, waterworks, a very good hospital, several warehouses, machine shops, a radio shack and tower, a commissary, a school, a church, a large club and mess-hall building, tennis courts, a golf course, and even a horse race track. The Fruit Company wanted to make life comfortable for its employees. Most houses, like the Popenoes', were screened, frame buildings constructed on "stilts" about seven feet above the ground for protection against flooding and to provide space underneath for doing the laundry, for a servant's room, for storage, and in the case of the labor camps, for cooking. It was a typical North American Fruit Company setup on the low coast of Central America: self-sufficient, and with sanitary conditions that were excellent compared to other parts of the isthmus. The incidence of malaria, smallpox, cholera, and hookworm had been drastically reduced; yellow fever had been eradicated. Tela, however, was an isolated community, provincial and totally cut off from the stimulating cultural activities the Popenoes had enjoyed in Washington.

Lancetilla Experiment Station near Tela, Honduras. The station was established by the United Fruit Company in 1926. Wilson Popenoe, its first director, was in charge of this station for several years. During the 1940s, it was considered the best arboretum of tropical fruit and forest trees in the Western Hemisphere. (*Courtesy of Baker Library, Harvard Graduate School of Business Administration, United Fruit Company Collection.*)

Lancetilla Experiment Station, 1926. Looking down the main path through the nursery area: on left, lemon seedlings to be used as rootstock for grafting; behind them, Para rubber seedlings; on the right, tung oil seedlings; a meteorological instrument shelter can be seen on the extreme right. (*Courtesy of Baker Library, Harvard Graduate School of Business Administration, United Fruit Company Collection.*)

There were several hundred acres of flat, alluvial land, some low hills, and a rain forest at Lancetilla. The climate was warm and humid, with an average rainfall of about 120 inches and no pronounced dry season. The station was readily accessible by rail car to the nearby port of Tela, and had fertile soil for nursery development—both important factors, since many thousands of fruit trees were to be distributed to other countries. Experiments were to be conducted that would benefit agriculture in Central America, including diversification of crops, introduction and testing of new crops, intensive study of soils, pest control, and trials involving several methods of propagation. Studies would be made of the growth and production of bananas, as well as research in banana diseases: Panama wilt disease, caused by a soil fungus, had been prevalent in Central America since the beginning of the century and required abandoning the infected land and planting bananas in new, uninfected areas; Sigatoka leaf spot, caused by a fungus that attacks the leaves of the plant, appeared suddenly in the region about 1935. These two blights would eventually destroy many thousands of acres of banana plantings throughout Central America and the West Indies.

Wilson Popenoe, at the age of thirty-four, was put in charge of the experiment station with the title of director of agricultural experiments for the United Fruit Company. He was impressed by the intense, rapid growth of the tropical vegetation around Lancetilla. He observed that once a horticulturist has lived and worked in the tropics, he will ever after become impatient when forced to grow plants in any other climate.

In the early days at Lancetilla, the United Fruit Company brought some experienced plantsmen to Honduras to assist Popenoe, including Alfred Butler, an old-school British horticulturist; A. N. Hartman from Romania to take charge of the banana experiments; and the Ecuadorian Jorge Benítez to work as an assistant horticulturist. Young research assistants arrived from Maine, Maryland, Alabama, and Missouri. Many other scientists collaborated in the development of Lancetilla, including David Fairchild, Dr. Vining C. Dunlap, Dr. Otto Reinking, Robert P. Armour, and Paul Allen, to name a few.

In addition to the banana, many other tropical crops were introduced and developed in this plant-introduction garden and experiment station, which eventually grew to some eight hundred acres under cultivation. Pineapples were obtained from several countries; many Asiatic fruits were brought in, including the durian, the litchi, the rambutan, the pulasan, and the langsat. The largest collection of Asiatic plants in the Neotropics was established at Lancetilla. Fruits introduced from other parts of the world included jaboticaba seed from Brazil and the akee fruit, *Blighia sapida,* from West Africa—an importation that was to have disastrous consequences for Wilson personally.

The largest planting of mangosteen, *Garcinia mangostana,* in the Western Hemisphere, about eight and one-half acres, was established with selected clones from Southeast Asia. A sizable collection of mangos and avocados was developed. Lancetilla introduced several high-yielding citrus varieties to Central America, including the Ponderosa lemon—which was the size of a grapefruit with the flavor of a lemon.

According to Wilson Popenoe, the outstanding contribution of Lancetilla to tropical agriculture in the Western Hemisphere was the introduction of the African oil palm, *Elaeis guineensis*; seeds were obtained from the Far East in 1926. Several years later, after careful observation, selected cultivars from Lancetilla were used to establish commercial plantings of African oil palm in Costa Rica and Honduras, which now amount to over fifty thousand acres. Modern factories have been put up to process this vegetable oil, which is used as a cooking oil, for margarine, and for soap manufacturing; the meal is an excellent source of protein for feeding animals.

Several acres of rubber, *Hevea brasiliensis,* were planted at Lancetilla in 1926 with selected seed from the Far East. Lancetilla played an important role during World War II, when high-yielding, disease-resistant rubber clones were distributed throughout Central America in cooperation with the U.S.D.A.

In 1928 an extensive collection of banana species, *Musa acuminata* and hybrids of Far Eastern origin, was established at Lancetilla. One unusual variety received from the Saigon Botanical Garden was the Valery, which was found to be resistant to Panama wilt disease. (Many years later, in the mid-1960s, the Valery gradually replaced the Gros Michel as the principal variety of the United Fruit banana plantings in Central America; in the 1980s it was superseded by a more wind-resistant variety called Grande Naine.)

The Tela Railroad Company maintained a laboratory at La Lima, Honduras, under the direction of Dr. Vining Dunlap, a plant pathologist from Cornell, to conduct soil and pathological studies. Wilson, in collaboration with Dr. Dunlap, worked on developing control methods to combat the second dreaded enemy of the banana, Sigatoka leaf spot disease. Through experimentation it was determined in 1935 that the most effective fungicidal spray for bananas was the Bordeaux mixture, a blend of copper sulfate, lime, and water. On one occasion, Wilson showed Samuel Zemurray, president of the United Fruit Company, how two acres of Gros Michel bananas that had been treated with the Bordeaux mixture were a beautiful, healthy green color, while several hundred acres of surrounding bananas had turned yellowish-brown and were dying, because of the ravages of Sigatoka. Zemurray was so favorably impressed by this island of green in a yellow sea that he ordered the United Fruit Company divisions in Central America to start spraying Bordeaux mixture, as soon as possible, on more than ten thousand acres of bananas. The results were highly satisfactory, which turned out to be feathers in the caps of both Popenoe and Dunlap. During the next twenty years Bordeaux mixture became the standard treatment for Sigatoka leaf spot disease on all the United Fruit banana plantations.

A large collection of abacá or Manila hemp, *Musa textilis,* a leading cordage material, was brought to Lancetilla from Panama in 1928. Selective breeding was carried out to improve the yield of fiber and fiber strength. Abacá is obtained from several species of wild plantains and bananas. With help from Lancetilla, extensive plantations of abacá were established in Central America during the war years of 1941 and 1942 to supply United States naval requirements of Manila hemp during a period when Far Eastern supplies were cut off by the Japanese.

Considerable work was carried out on tropical timbers including mahogany (Central American and African); teak (*Tectona grandis*); Central American cedar and cypress; primavera (white mahogany); Burmese rosewood; sandalwood; tropical pines; and *Eucalyptus deglupta,* a tree that grows very rapidly in the lowlands of Central America and is an excellent source of pulp and construction lumber. Spacing tests were carried out to evaluate the growth rate and potential production of the most important tropical timber trees. Asiatic timber bamboos were planted. In many regions of the Far East bamboo plays an important role in home construction and furniture making.

Eventually some eight hundred varieties of plants of economic importance from all over the tropics were established at Lancetilla. Experimental plantings were made of cacao; chicle; tung oil trees of the genus *Aleurites*; forage grasses; sugar cane; spices; edible nuts; essential oil crops including citronella, lemon grass, vetiver, and ylang-ylang.

There was a collection of medicinal plants including *Dioscorea* (source of cortisone); *Derris* and *Lonchocarpus* (sources of rotenone insecticide); *coca*; *Rauwolfia*; and *Aloe.*

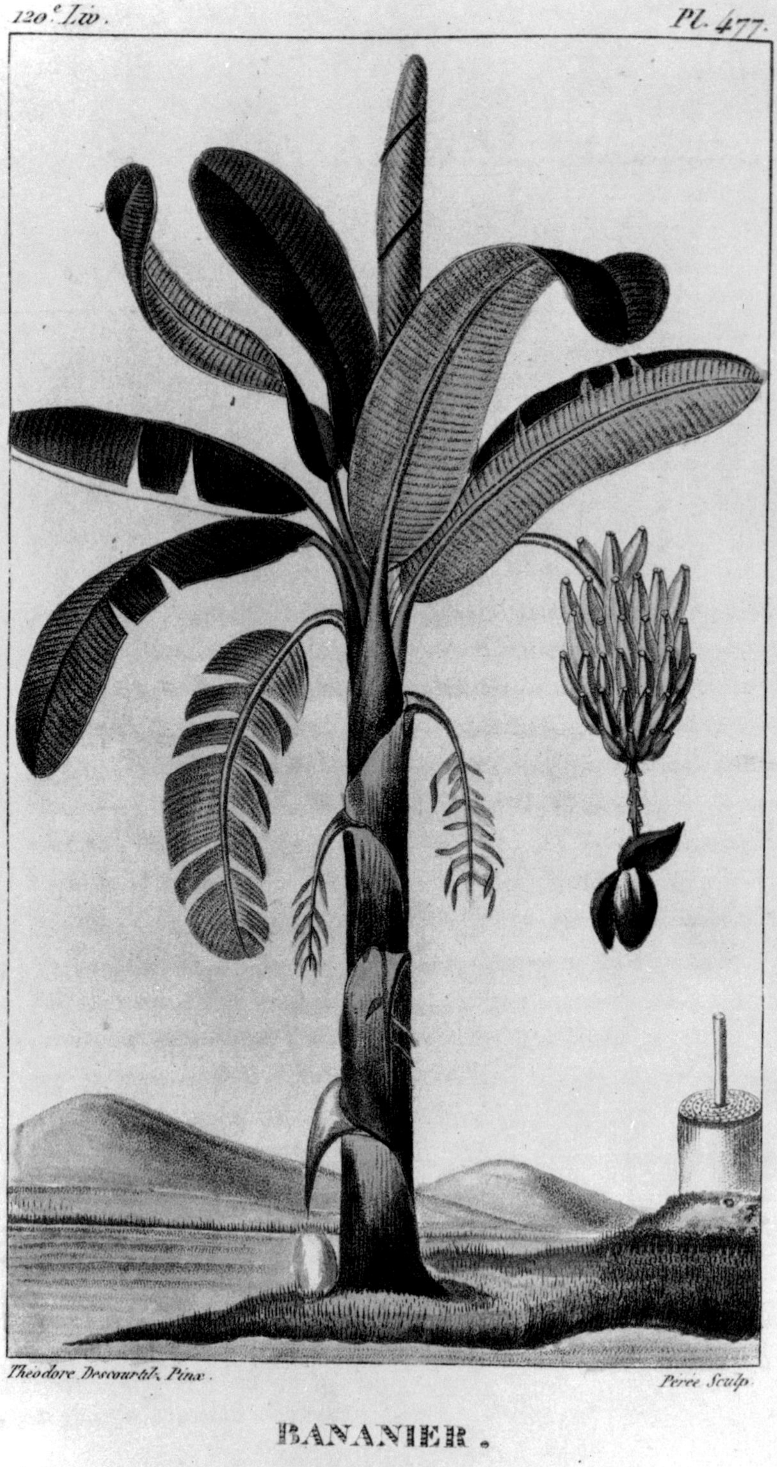

The banana, *Musa acuminata.* (*From M. E. Descourtilz,* Flore Pittoresque et Médicale des Antilles, *2d ed. 1833. Courtesy of Botany Libraries, Harvard University, Cambridge, Mass.*)

Several species of coffee were established including *Coffea arabica, C. robusta, C. liberica,* and *C. stenophylla.* It was hoped types of coffee might be found that could be cultivated successfully at low elevations in the coastal regions of Central America. Central American coffee is generally grown at higher elevations of fifteen hundred to five thousand feet.

In addition to serving as an experiment station for tropical plant introductions, Lancetilla was a base for scientific investigations in botany, zoology, ornithology, and entomology. Dr. Paul Standley stayed there during 1927 and 1928, investigated the flora of the region, and in 1931 published a noteworthy book entitled *Flora of the Lancetilla Valley, Honduras.* In 1928 a serpentarium or "snake farm" was established in cooperation with Harvard University, the Antivenin Institute of America, and the Mulford Drug Company of Glenolden, Pennsylvania, which lasted for several years. Venom from which antitoxins are made was extracted from dreaded serpents such as the fer-de-lance "barba amarilla" and the Central American rattlesnake "culebra de cascabel." Several world-renowned ornithologists visited Lancetilla to study tropical bird life. James L. Peters published a bulletin on the results of his observations.

Dorothy Popenoe welcomed the move to Honduras. She had a very inquisitive mind and not only learned Spanish quickly but also soon mastered the rudiments of the Carib Indian tongue. She developed a keen interest in the archaeology of Central America. In 1927 after traveling several days on muleback through rugged terrain in Honduras with only the faithful Jorge Benítez as a companion, she visited the prehistoric mountain fortress of Tenampua; she then prepared a scholarly report which was published by the government of Honduras. Dorothy and Wilson together wrote several articles on the history and archaeology of Honduras. One called "The Human Background of Lancetilla" came out in the August 1931 issue of *Unifruitco Magazine.* In this report, they showed their affection for and knowledge about Honduras. They described how the Spanish conquistador, Cristóbal de Olid, veteran of the arduous campaign in Mexico that resulted in the fall of Montezuma's empire, was dispatched by Hernán Cortés to undertake the conquest of Honduras. The Spaniards landed at the bay of Tela in April 1524 and established a settlement called Triunfo, near Lancetilla, which was later abandoned. Although they fought among themselves, the Spanish invaders managed to conquer the native inhabitants of Honduras in a cruel and ruthless manner. At the same time they put down in writing much of what they saw and

Dorothy Hughes Popenoe, 1899–1932. (*Courtesy of Hunt Institute for Botanical Documentation, Carnegie Mellon University, Pittsburgh, Pa.*)

Dorothy Popenoe, a proficient archaeologist, excavating a pre-Colombian cemetery at Playa de los Muertos, Ulua River valley, Honduras, 1929. (*Courtesy of Peabody Museum, Harvard University, Cambridge, Mass.*)

heard which formed a background for subsequent archaeological studies, such as Wilson and Dorothy's, which attempted to reconstruct the story of the pre-Colombian peoples of the region. During the clearing of the land to develop the Lancetilla Agricultural Experiment Station, the laborers uncovered numerous obsidian artifacts, stone hatchets and corn grinders, effigy jars, arrow heads, jade amulets, and pottery—relics of the early inhabitants of the Lancetilla valley. Wilson and Dorothy accurately described these remains of the Mayan past in an interesting essay which was illustrated by Dorothy.

Dorothy was intensely interested in Mayan art and archaeology. She spent several weeks in 1928 and 1929 excavating a pre-Colombian cemetery next to a remote beach in Honduras known as Playa de los Muertos on the Ulua River. A gallant, adventurous spirit, she never hesitated to camp alone in that desolate region, notorious at the time for its bandits and cutthroats. Utilizing her literary and artistic skills, she wrote a scholarly report about the pottery and skeletons of Playa de los Muertos, illustrated by her drawings. It was published in 1934 in *Maya Research* by the Alma Egan Hyatt Foundation, New York.

The akee, *Blighia sapida.* The akee from tropical West Africa was brought to the Americas in the days of the slave trade. The outer pod of the oily fruit is straw colored to bright red; the inner flesh is yellowish. The unripe, uncooked akee is highly poisonous—the toxic principle in it is hypoglycine—and is believed to have caused the death of Dorothy Popenoe, Wilson's first wife, in Honduras in 1932. (*Photograph by Wilson Popenoe.*)

At the "Beach of the Dead" she had uncovered the remains of some fourteen bodies as well as a varied collection of artifacts including pottery, shells, beads, and ornaments, some of which gave clues concerning the distant past of the Mayan race.

Dorothy was a sensitive, perceptive writer. On one occasion, in a letter, she described vividly a Central American jungle scene:

> We entered the real tropical jungle where the bright blue sky was reflected in patches through the tangle onto deep green water; it seemed like a river of mother of pearl. On all sides was this dense struggling, sprawling, climbing mass of vegetation. The jungle seemed to be giving birth continually, guilty of mad, ungoverned spawning. Death comes to the things of the jungle not so much from extraneous accident as from sheer presence of birth.

When she first came to Honduras, Dorothy had a pink and white complexion and a youthful appearance. By 1932, however, having given birth to five children in just a few years while continuing to work on archaeology in Honduras and the reconstruction of a seventeenth-century house in Antigua, Guatemala, she seemed tired. In

December of that year, a very tragic event took place in Lancetilla. Dorothy Popenoe ate an unripe, uncooked akee fruit which is said to have poisoned her. Seized by violent convulsions, in great pain she was rushed to the United Fruit hospital at Tela, where an emergency operation was performed. It was too late to save her life. Wilson's talented young wife died in the hospital at the age of thirty-three. At the time, he was away on a trip. Dorothy was buried on a hillside above the Lancetilla Agricultural Experiment Station.

At the Peabody Museum of Archaeology and Ethnology, Harvard University, there are some 150 photographs taken by Dorothy Popenoe of various archaeological sites in Honduras and other parts of Central America. There is also a collection of ceramic artifacts from Honduras, excavated by Dorothy Popenoe—some of which were presented to the Peabody Museum by Dorothy during her lifetime, and the remainder donated at a later date by Wilson, in her memory.

Saddened by Dorothy's untimely death, Wilson was in a quandary as to what he should do. He now had five young children on his hands, including a baby less than a year old. Charles H. Popenoe, a first cousin of Wilson, had just died unexpectedly of a heart attack, leaving a twenty-five-year-old widow named Edith, who lived in Silver Spring, Maryland. She and Wilson made a most convenient arrangement in 1933: Edith took his children to live with her and her own two small children in Maryland. He contributed financially to support Edith's large household. Wilson was thus able to continue working for the United Fruit Company in Central America. This harmonious and fortuitous family relationship with Edith lasted for nine years. Edith, while raising all those children, took a master's degree in psychology and eventually became supervisor of pupil services for the Montgomery County (Maryland) school system, heading a large staff. Many years later she married Wilson's old friend Knowles Ryerson, who had become a widower in 1968.

Under Wilson's capable guidance, Lancetilla grew in importance. He stayed on as director of the station for several years but then gradually moved on to other work with United Fruit, mainly as a technical advisor on growing bananas and other crops at the company's divisions in Central America and the Caribbean.

Lancetilla was in its prime during the 1930s and 1940s. Hundreds of thousands of plants were propagated and distributed to institutions, individuals, and various governmental agencies in the American tropics. Knowles Ryerson in his taped memoirs stated that in 1934 Lancetilla was "the best arboretum of tropical fruit and forest trees in the

Western Hemisphere." By 1965, however, the United Fruit Company started to look for some scientific organization to take over the administration of the experiment station. For several years the place went downhill, until in 1974 the company transferred it to the government of Honduras. The situation at Lancetilla has subsequently improved, a turn of events that would have pleased Wilson Popenoe. On many occasions he and David Fairchild had discussed the importance (and unfortunate lack) of continuity in some botanic gardens and arboreta in Latin America.

In 1974 a new name was given to the agricultural experiment station at Lancetilla: it is now called "Dr. Wilson Popenoe Botanical Garden" (Jardín Botánico Dr. Wilson Popenoe) in honor of the man who guided its destiny in the early years of its development.

From 1933 to 1940 Wilson spent less time at Lancetilla. He had been named chief agronomist and director of tropical research for the United Fruit Company, and as such did a lot of traveling—checking on the condition of the company's 125,000 acres of bananas in Honduras, Guatemala, Costa Rica, Panama, and Colombia. He also visited Jamaica from time to time, where he gave advice to cooperating banana growers concerning the control of leaf spot (Sigatoka), which was becoming a major problem. His friends called him "The Traveling Farmer," since he was constantly moving about giving technical advice about bananas. He pointed out to the government agricultural officials and private growers in Jamaica that you should not spray the bananas a few times with Bordeaux mixture and then stop and forget about it. It was important to keep on spraying persistently. To be successful, he advocated the biblical injunction "Be not weary in well doing."

In 1939 he wrote an unpublished report about Jamaica, which contained the following observations:

> For many years now, my work has kept me principally in the Spanish-American republics. But annually I have visited Jamaica, sometimes to spend a week, sometimes to spend several months. And each time I have come, I have been struck by the contrast between Anglo-Saxon tropical America and Latin tropical America. The two are very, very different. Physically, Jamaica is not unlike the eastern part of Cuba, or Santo Domingo or Puerto Rico. But there the likeness ends. When you step ashore in Kingston, fresh from any one of the Latin republics, you are struck immediately by the architecture of the place. You miss the patio houses, outer walls flush with the sidewalk, to which you have been accustomed.

> If you are an American, the older buildings, particularly the churches, remind you more of Colonial Virginia than they do of anything else—and they should, since they were built at the same time and by the same race. . . . The hotels in Jamaica are cleaner; the food, while wholesome, may not be as interesting. There is rather too much of boiled fish and boiled beef, while as for the vegetables, there may be only three—as Irvin Cobb once put it—boiled potatoes, boiled cabbage, and a second helping of boiled potatoes. . . . In Jamaica it is comfortable, but not as colorful as Guatemala. After I have been in Jamaica a few weeks, I begin to think: "How pleasant it would be this morning to wake up hearing the military band playing as they march the soldiers out to drill. Playing rather badly, but playing with a bravado which stirs one's blood. And how pleasant it would be to have an Indian boy bring in my morning coffee. Maybe he would spill part of it on the bedspread, but we would laugh over that together, and he would enquire with genuine solicitude about my night's rest." Yes, Jamaica is a beautiful island, preferred by many to our more romantic, if at times less clean, Latin countries. There is room, I believe, for both points of view.

Wilson believed strongly that in the smaller countries in Latin America and the Caribbean region, which depend on agriculture, schools should stress agriculture in their teaching, rather than train young men for city jobs. He was in agreement with Carlos Figueroa, former dean of the College of Agriculture at Mayaguez in Puerto Rico, who stated: "You Americans have made a great mistake. You came to this island of ours, Puerto Rico, an island where agriculture is the only hope, and you built high schools all over the place. You educated our youngsters *away from the land.* For when they had a high school education, they all wanted to live in town and have white-collar jobs."

This was the point of view that was to influence Wilson two years later, in 1941, to take on the challenging job of becoming the founding director of Escuela Agrícola Panamericana in Zamorano, Honduras.

Chapter X

Antigua, Guatemala, 1930–1936: Casa Popenoe

In the late 1920s, while living in the warm tropics at Tela, Honduras, Wilson and Dorothy used to visit the cool, temperate highlands of Guatemala from time to time for a refreshing change of climate. Irresistibly attracted to Antigua, they both fell in love with the ruined city and started to look for a house they might buy and restore. After looking at fifteen or twenty, they found the one they wanted on Primera Avenida, a street known in olden times as Calle de la Nobleza (Street of the Nobility)—where the homes of the prosperous used to flank it on either side. They bought the property in 1929 for the bargain price of two thousand dollars. The annual taxes were four dollars and eighty-five cents.

In his booklet entitled *La Casa del Oidor,* Wilson described the condition of the house:

> When we purchased it in 1929 from don Ciriaco Peralta it was in a state of almost complete ruin; most of the roof was gone, there was only one door left in the entire house, the corridors had disappeared completely, and the brick floors were gone except for a small area in the sala [main hall]. To carry out the restoration, doors and shutters were purchased from old houses in Antigua; the handsome window grilles and the large door to the sala were copied from another grand house of the period; the roof had to be rebuilt completely; and new brick floor laid. It was interesting to find that no nails had been used in the original overhead work; the trusses were tied together with rawhide thongs.

The debris left by the earthquake of 1773 had been pushed into the courtyard and garden. A tall Capuchino cypress tree, planted about 1865, dominated the ruins. In recent years the house had become known as the Casa del Capuchino because of this prominent, tall tree.

In 1930 Wilson was transferred to Guatemala City to organize the Technical Service of Agricultural Cooperation (Servicio Técnico de Cooperación Agrícola) for the United Fruit Company, with the ob-

Antigua, Guatemala, in 1839. (*From John Lloyd Stephens,* Incidents of Travel in Central America, Chiapas and Yucatan, *with illustrations by Frederick Catherwood, London, 1854.*)

jective of introducing new crops to Guatemala. He and Dorothy started at once the reconstruction of the house which had originally been called La Casa del Oidor (The Judge's House) in the seventeenth century. (Wilson always gave credit to Dorothy, rather than to himself, for this ambitious project.) The house was built about 1634 by don Luis de las Infantas y Mendoza, a wealthy doctor of law from the famous Spanish University of Salamanca, who had a legal practice in Sevilla. With his young bride and six servants don Luis arrived from Spain at Santiago de los Caballeros de Guatemala (now Antigua) to assume the important and lucrative position of *oidor* (judge of the chancery) of the Real Audiencia (Royal Supreme Court) which had jurisdiction over the Captaincy General of Guatemala including the Mexican state of Chiapas and most of the region now known as Central America—Guatemala, Honduras, El Salvador, Nicaragua, Costa Rica, and part of Panama. The position of oidor in Guatemala was a bonanza for don Luis for a few years—highly remunerative because of the large bribes he received. After a person had been found guilty by other judges, the oidor would impose the penalty or fine. Thomas Gage, a British Dominican friar and author who lived in Antigua in the 1630s, knew don Luis de las Infantas y Mendoza. Highly critical of the dishonesty of Spanish authorities in the New World, Gage mentioned the financial dealings of don Luis in his classic *The English American . . . or, A New Survey of the West India's*:

Believe It or Not —BY RIPLEY

(*Reprinted by permission of UFS, Inc.*)

> The pension which the King alloweth to every judge of Chancery is four thousand ducats yearly. . . . Yet what besides they get by bribes and trading is so much, that I have heard a judge himself Don Luis de las Infantas say that though a judge's place at Mexico and Lima be more honorable, yet none more profitable than Guatemala.

Influenced by his background in Sevilla and sparing no expense, don Luis built a magnificent house which featured the architecture of Moorish Spain. Three centuries later, although in poor condition, the house's handsome stone arches, the remaining huge cedar door, and beautifully turned shutters exemplified the Moorish influence of Sevilla, Córdoba, and Granada. The Moors, being Moslems, were forbidden to make use of florid designs so the carpenters and masons restricted themselves to geometric patterns.

Main patio of the Casa del Oidor, Antigua, Guatemala, before reconstruction, 1929. The house became the Casa Popenoe. (*Courtesy of Hugh Popenoe.*)

Don Luis made the mistake of quarreling with the governor of Guatemala, an error that cost him his job. He was obliged to take a lowly position in Mexico in 1640 as a minor tax assessor and is said to have died in poverty. The Casa del Oidor passed through several hands, seemingly bringing bad financial luck to its owners. For example, in 1738 it was owned by a family named Herrera who, in a mood of generosity and religious fervor, promised to contribute four hundred pesos to the nearby monastery of Santo Domingo. This turned out to be a disastrous pledge. When the Herreras, having suffered financial reverses, were unable to make the payments they had promised, the monastery sued them. The court ordered the house to be sold in 1755. It was purchased for a relatively low price by a captain of the guards, Diego de Guerra. After court and legal expenses had been taken care of, there was nothing left for the Herrera family; virtually penniless, they were ejected forthwith from the premises.

By this time, Santiago de los Caballeros de Guatemala was a thriving capital, inferior in importance in the New World only to the capital of Mexico and Lima, Peru. There was a magnificent cathedral, a stately palace of the Captains General, some eighty church edifices, numerous monasteries, cloisters, public buildings, and houses, built of solid masonry, many with vaulted roofs and walls three to four feet thick. The

Jorge M. Benítez. For over fifty years Benítez was a loyal co-worker and friend of Wilson Popenoe. An excellent horticulturist, Benítez was also in charge of the reconstruction of the Casa Popenoe in Antigua, Guatemala, from 1930 to 1935. (*Courtesy of Hunt Institute for Botanical Documentation, Carnegie Mellon University, Pittsburgh, Pa.*)

University of San Carlos was the greatest seat of learning in Central America. It was a picturesque city, where Spanish colonial life was flourishing. The population had grown to fifty thousand.

On July 29, 1773, at 3:30 P.M., a sharp tremor shook the Guatemalan capital. As usual in this earthquake-prone area, everyone rushed out onto the streets; they had been experiencing many tremors in recent months. There was a period of calm for about ten minutes, then the most dreadful earthquake in the city's history struck. The earth heaved so violently many screaming, panic-stricken people could not stay on their feet. Within two minutes, almost all the buildings in the city had collapsed, thrown to the ground in a heap of mortar, wrecked beyond repair. Within a few days, the atmosphere became pestilential as human and animal corpses corrupted the air. Overwhelmed by the disaster, the citizens of Santiago petitioned the king

Courtyard of Casa Popenoe after reconstruction. In keeping with the Moorish influence notable in the colonial architecture of Antigua, the elegant houses had unimpressive exteriors, but charming inner courts and gardens. (*Courtesy of Nixon Smiley.*)

of Spain for permission to move the capital of Guatemala to an area less subject to earthquakes. The king reluctantly consented; in 1777 the capital was moved to the Valle de las Vacas, some twenty-six miles to the east. Virtually abandoned, Santiago became known as Antigua, or the old city. All the Antigueños who considered themselves worth anything moved to the new capital (Guatemala City), leaving shattered Antigua to rest in peace in the shadows of its volcanoes. The ruined churches, government buildings, and homes were stripped of movable valuables—which were loaded onto the backs of mules or Indians and transported to the new city. The poor people left behind, mostly Indians, moved into the abandoned ruins, one of which was the Casa del Oidor. Captain Guerra's son subsequently sold the

THE ARCH OF SANTA CATARINA, WITH THE DOME OF MERCED IN THE DISTANCE

Dorothy Hughes Popenoe was a talented artist, as well as an archaeologist. This is one of her sketches of Antigua. (*Reprinted by permission of the publishers from* Santiago de los Caballeros de Guatemala *by Dorothy H. Popenoe, Cambridge, Mass.: Harvard University Press, Copyright © 1933 by the President and Fellows of Harvard College.*)

property which passed through several more hands until purchased 150 years later by Wilson and Dorothy Popenoe.

In 1932 Wilson was transferred back to Tela. The restoration of the Casa Popenoe (as it was now known) was continued under the supervision of his faithful and competent horticultural associate, Jorge Benítez, whom he referred to as his "Ecuadorian man Friday." Every effort was made to bring the colonial past back to life. The master

Louis Adamic, author of *The House in Antigua,* with Wilson Popenoe, Casa Popenoe, 1936. (*Courtesy of Hugh Popenoe.*)

craftsmen involved in the repairs were Arturo Santis, a mason; and Julio Gómez, a carpenter. Santis was paid eighty-five cents a day for several years; his wages were then increased to a dollar twenty-five a day. Wilson asked Santis, after the raise, if he was faring better than before. The mason replied, "I am making more money, it is true; but when I was earning eighty-five cents a day I could buy fourteen tortillas

Master bedroom, Casa Popenoe. (*Courtesy of Hunt Institute for Botanical Documentation, Carnegie Mellon University, Pittsburgh, Pa.*)

for one cent. Today I can only buy four tortillas for a cent. So you see I am not better off."

In late 1932, shortly before her death, Dorothy Popenoe wrote a fascinating booklet entitled *Santiago de los Caballeros de Guatemala,* illustrated by her own drawings. It was published posthumously by Harvard University Press in 1933. In this brief, well-written work, she described vividly the conquest of Guatemala, and the history of Antigua including its inhabitants, churches, earthquakes, and numerous ruins.

By 1936 the restoration of the Casa Popenoe was nearly finished. The author Louis Adamic, well-known for his book *The Native's Return,* visited Antigua and stayed with Wilson for four weeks. He then wrote *The House in Antigua,* which was published in 1937 by Harper and Brothers. Adamic's book, dedicated to the memory of Dorothy Popenoe, caught the spirit behind the restoration as he described the romantic history of the Casa del Oidor during three centuries. Adamic's book was widely read in the United States. Many tourists who visited Guatemala wanted to go to Antigua to see the Casa Popenoe. A plan was worked out—with no financial reward for Wilson—so that guides could escort tourists to the house. Wilson's venerable Indian housekeeper, María, would then show them around.

The house thus became not only a popular tourist attraction, but a cultural link with Antigua's colonial past.

Wilson insisted on keeping everything in the house just as it had been in the eighteenth century: no telephone, no electricity, no contact with the modern world. His favorite mahogany desk was made in 1700. One night fifty years ago, as a guest at the Casa Popenoe, the author became aware of its colonial mystique. We had just arrived late in the evening following a visit to a cinchona plantation. Since I had never seen the house before, Wilson, by candlelight, showed me his remarkable collection of authentic, old Antiguan furniture including tables, chairs, chests, benches, cabinets, and beds, mostly made of Spanish cedar—which had a unique, musty, woody aroma. The portraits adorning the walls, dimly visible, portrayed many notable Spaniards who had taken part in the colonization of Guatemala, including soldiers, poets, teachers, priests, and a nun. The smoke of three centuries was still on the walls of the domed kitchen. María was adding to it as she grilled our dinner over charcoal, the same fuel that had been used in that same stove in colonial times. Although it was 1940, it seemed like 1773.

Chapter XI

Guatemala, 1940: Cinchona (Quinine)

It has been estimated that over the course of history malaria has killed more humans than all wars and plagues combined. Countless millions who survived this dread disease suffered in misery from chills and fevers, chattering of the teeth, nausea, exhaustion, and permanent damage to the brain and liver.

In the seventeenth century it was discovered that a medicine obtained from the bitter bark of some South American trees known as *quina quina* could cure malaria. These were handsome evergreen trees of the genus *Cinchona,* native to the Andean highlands, mainly in Peru, Bolivia, Ecuador, and Venezuela.

Although the curative effects of *quina quina* bark had been known to the Peruvian Indians for centuries prior to the arrival of the Spanish conquistadores, the earliest reputed use of the drug to cure fevers in Europeans occurred in 1638. In that year Ana, the countess of Chinchón, living in Lima, Peru, is said to have been cured of tertian fever thanks to a packet of powdered Peruvian *quina quina* bark, after all other remedies had failed. Upon her recovery, the countess was so grateful she caused large quantities of the bark to be collected and given to those sick with the fever. The medicine became known as *Polvo de la Condesa,* or the Countess' Powder; also as Chinchona bark, Peruvian bark, and Jesuit's bark—since the Jesuit fathers distributed the febrifuge throughout Latin America and Europe. Because of an error in spelling, the fever-bark tree was named *Cinchona* (not *Chinchona*) by Linnaeus, the noted Swedish botanist.

A great demand developed for cinchona bark. At first, the South American supply seemed inexhaustible, but as more and more trees were cut down and their bark stripped off, it became evident that cinchona plantations would have to be established to provide adequate supplies of the precious bark for future generations. Several governments in South America, however, to protect their patrimony, enacted strict laws, including the death penalty, to prevent the export (or

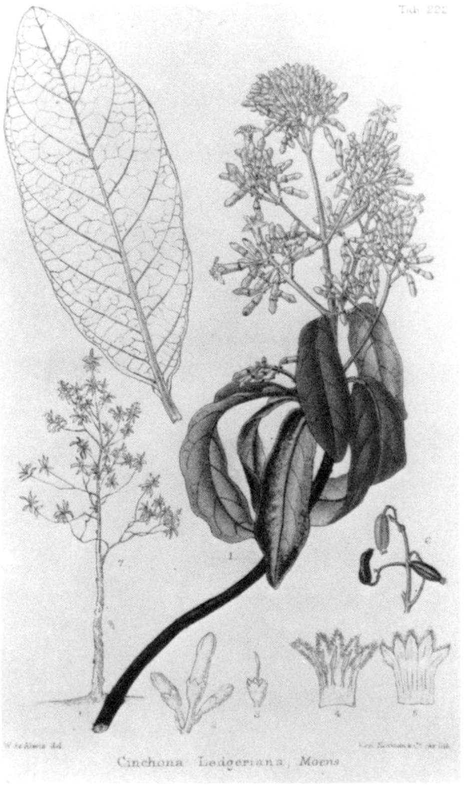

Quinine is extracted from the bark of the cinchona tree. (*From* Journal of Botany, *vol. 19, London, 1881. Courtesy of Botany Libraries, Harvard University, Cambridge, Mass.*)

smuggling out) of cinchona seeds or cinchona planting material of any kind.

In 1863, an English trader living in Bolivia named Charles Ledger managed to smuggle several pounds of selected cinchona seed out of Bolivia, which were sold to Dutch and British interests. These seeds were destined to become the main basis of the large, successful Dutch plantations in Java. The high-yielding species was named *Cinchona ledgeriana* in recognition of Ledger's efforts; as a reward for his services to England, he was knighted by Queen Victoria. Cinchona plantations were also established by the British in India at about the same time. But the Dutch in Java and Sumatra dominated the market, establishing a monopoly which by World War II controlled 90 percent of the world supply of cinchona bark. By the 1930s the production of the bark in South America had fallen to insignificant levels.

In 1932 Merck & Co., Inc. of Rahway, New Jersey, became interested in developing cinchona in the Western Hemisphere in cooperation with the U.S. State Department. It was deemed expedient to build up supplies of cinchona bark for quinine in order to become

Map of Guatemala showing cinchona (quinine)-growing fincas, 1940. (*From Frederic Rosengarten, Jr.,* History of the Cinchona Project of Merck & Co., Inc., and Experimental Plantations, Inc., 1934–1943. *Merck & Co., Inc., Rahway, New Jersey, 1944.*)

independent of the Dutch quinine monopoly. Following considerable study and consultations with Wilson Popenoe, Guatemala was selected as the most favorable Latin American country in which to start growing cinchona. Favorable factors included the support of the Guatemalan government, fertile soils, a desirable climate, a wide choice of altitudes and rainfall, and good possibilities for collaboration with local planters.

Wilson Popenoe recommended Antigua, Guatemala, as a logical site to begin seedbed and nursery work, under the supervision of Jorge Benítez, a trained horticulturist. Land for the cinchona nursery was

Cinchona succirubra at Finca El Naranjo, Chicacao, Guatemala. *C. succirubra* is a vigorous species with a low quinine content. From 1942 to 1944 the U.S. government financed a crash program in Guatemala to produce millions of *C. succirubra* seedlings which were supposed to be harvested at two to three years of age to produce quinine for the war effort. The program was subsequently abandoned. (*Photograph by the author.*)

rented from Pedro Cofiño, an old friend of Dr. Popenoe, at Finca La Retana, Antigua.

The Pacific coast of Guatemala was selected for trial cinchona plantings, to be carried out at existing coffee plantations. The first cooperating planters, well known to Wilson Popenoe and recommended by him, were Gordon Smith at Finca Mocá, Guatalón, in the department of Suchitepéquez (to be joined a few years later by his sons Owen and John G.); L. Lind Pettersen at Finca El Zapote, above Escuintla; and Percy Davies at Finca Helvetia, in the department of Retalhuleu. (Some thirty years later this same Finca Helvetia, under different ownership, was used by anti-Castro Cubans as a training site for their ill-fated invasion of Cuba known as "The Bay of Pigs.")

Dr. Popenoe recommended that additional experimental plantings of cinchona should be made at other Guatemalan locations, including Finca Samac, Cobán, and Finca El Porvenir, San Marcos. As a matter of historical interest, in 1883 the Guatemalan government had already planted cinchona at El Porvenir with seeds of *Cinchona succirubra* from Ceylon (Sri Lanka); by the late 1930s thousands of vigorous old *succirubra* trees were growing at El Porvenir in semi-abandonment. *Succirubra* bark, however, has a low quinine content of only about 2 percent, whereas high-yielding *C. ledgeriana* bark, with a quinine content of 7 to 14 percent, was needed for the new project in Guatemala. Cinchona bark contains four main alkaloids: cinchonine, cinchonidine, quinine, and quinidine. All are useful as antimalarials, and quinidine sulfate is also utilized in medicine to regulate the heartbeat, a therapy for certain cardiac arrhythmias.

In 1933, Colonel Victor Ruehl made a trip to Java for Merck & Co., Inc., where he obtained some high-yielding *Cinchona ledgeriana* seed for planting in Guatemala. For several years, scions of selected *ledgeriana* clones (from Ruehl's seeds) were grafted onto *succirubra* rootstock (of El Porvenir origin) in the Guatemalan nurseries.

We have noted how Sir Charles Ledger smuggled cinchona seeds out of Bolivia in 1863, which led to the establishment several years later of the great Dutch cinchona plantations in the Far East. A somewhat similar event took place in 1876, when Sir Henry Wickham "took" some seventy thousand rubber (*Hevea brasiliensis*) seeds from the Amazon in Brazil to be grown at Kew in England. The seedling plants were subsequently transported to the Far East and used to develop the huge rubber estates in British Malaya and the Dutch East Indies. Both of these noteworthy smuggling operations were highly successful at the expense of Latin America.

Many decades later, a somewhat similar act of smuggling may have taken place, but this time in the opposite direction, that is, from east to west. An American soldier of fortune, who shall remain nameless, is said to have visited the Dutch East Indies in the 1930s, looking for selected cinchona seed. The Dutch government, protecting its quinine monopoly, strictly prohibited the export of any kind of cinchona seeds or cinchona plant material from Java and Sumatra. Nevertheless, this intrepid adventurer was determined to obtain high-yielding *Cinchona ledgeriana* seed from a Dutch plantation which happened to be owned by a wealthy widow. It was well known that this proud lady would

Finca Panamá, Guatalón, Guatemala, June 1940. Left to right: the author, F. W. Owen Smith, and John G. Smith. In background, *Cinchona succirubra* plants which have just been used as rootstock for grafting of *C. ledgeriana* scions, high in quinine content. (*Photograph by the author.*)

refuse to sell her cinchona seed to any foreigner, regardless of price. So what could he do? He devised an ingenious plan. A charming, if elderly gallant, he decided to call on the widow. A tireless Casanova, he got to know her quite well. After flattering her outrageously, one day he asked her to take him for a walk to show him her cinchona plantation. Rain was threatening, so he took along an umbrella. They sat down under a high-yielding *ledgeriana* tree. He opened his umbrella and laid it on the ground, face up. While making energetic, amorous advances to the unsuspecting widow, he managed somehow to shake the tree. Hundreds of tiny cinchona seeds fell into the open umbrella, and were planted a few months later in the Western Hemisphere. Can this be true? Quién sabe?

In May 1939, following a recommendation of Wilson Popenoe, Merck & Co., Inc., through its subsidiary Experimental Plantations, Inc., purchased Finca El Naranjo, Chicacao, Suchitepéquez, some 120 miles west of Guatemala City on the western slopes of the volcano Santa María, with an altitude range of 3,800 to 7,500 feet. This plantation was to be the Central American cinchona headquarters for Merck for many years. Dr. Popenoe was much impressed by the volcanic soils of the southwest coast of Guatemala—a region he considered to be among the most fertile in the tropics of the world.

In April 1940, Merck & Co., Inc. made an arrangement with the United Fruit Company whereby Wilson Popenoe was enabled to take charge of Merck's cinchona project in Guatemala for a period of five months. At the time the United States was faced with the threat of fighting a war in the Pacific which could result in the cutting off of the traditional supplies of cinchona bark from the Dutch East Indies, thereby causing a critical shortage of quinine as an antimalarial for our armed forces.

Popenoe immediately organized an extensive bark-sampling program to determine the quinine content of hundreds of *Cinchona ledgeriana* trees in Guatemala—to select clones for future asexual propagation. He likewise installed for the first time a conventional system of labeling and numbering trees for the fincas growing cinchona in Guatemala. (The labeling system would have been more effective, however, if the Indian women at the plantations hadn't repeatedly removed the handsome copper labels from the trees to use as jewelry.)

Wilson pointed out the importance of training young Guatemalans to carry out nursery work, such as grafting, as well as field work in the cinchona plantings. He and his assistant (the author) visited many Guatemalan coffee fincas in 1940, distributing cinchona seeds and grafted plants, while encouraging Guatemalans to grow cinchona trees—to produce more quinine in Latin America and to diversify their crops. In August 1940 we made a trip to Finca El Porvenir, starting early in the morning from Casa Popenoe in Antigua. However, it was not as early as planned since Juan, the chauffeur, arrived two hours late. "Que pasó?" (What happened?) asked Dr. Popenoe. The chauffeur answered, "Es que me agarró el sueño" (sleep got the best of me). Wilson, always in good humor, agreed that was as good an excuse as any, so we started off.

In a dilapidated Ford sedan we made our way along very rough roads (it was the rainy season) from cool Antigua down to the hot low coast. Near the town of Patulul the car got stuck in the mud, so

Wilson Popenoe with Lake Atitlán, Guatemala, in the background, 1940. (*Photograph by the author.*)

deep in the quagmire the chains on the tires were of no use. We were pulled out by a yoke of oxen—an everyday occurrence in those days. From San Felipe, Retalhuleu, we started up the former roadbed of the abandoned railroad known as "Ferrocarril de los Altos" which had been converted into a road for automobiles. About half way up to Quetzaltenango, we came to a tunnel. Juan, who was highly superstitious, abruptly stopped the car and refused to proceed. He asserted that we should not go through the tunnel, since it would bring all of us bad luck, including his car, so we had better turn back. We stayed at the entrance to the tunnel for quite a while until Dr. Popenoe hit upon a solution to the impasse: Juan should get out of the car, walk above the tunnel on the side of the mountain, and join us at the other end. Muttering under his breath, Juan reluctantly agreed. The author then drove the car through the tunnel and we continued on our way, reaching the town of San Marcos at nightfall.

Drying cinchona bark, Loja, Ecuador, 1941. In 1943 Wilson Popenoe traveled through the cinchona-producing regions of Ecuador, Peru, and Bolivia, to help organize the procurement of cinchona bark for the U.S. armed forces, which were threatened by a scarcity of antimalarials. (*Courtesy of Hunt Institute for Botanical Documentation, Carnegie Mellon University, Pittsburgh, Pa.*)

The hotel was filthy, but we were thankful to have a roof overhead since it was raining heavily. It was cocktail time, so Dr. Popenoe produced a small bottle of Scotch and we had a drink. He quoted Dr. McPhail, the United Fruit Company physician: "One drink in the tropics is good for you, and two won't do you any harm."

The hotel didn't have a bathroom, just a stinking privy in the backyard. In due course, Dr. Popenoe smiled as he lit one of his favorite Havana cigars. Always the optimist, he observed, "With one of these fine cigars, I could go anywhere," and stepped resolutely outside.

The next morning we proceeded to Finca El Porvenir, a beautiful coffee plantation not far from the Mexican border on the Pacific slopes of the volcano Tajamulco. At the time, this was one of the largest coffee-producing plantations in Central America. We were greeted politely by the German manager. El Porvenir belonged to the Capco, a German company that was subsequently nationalized by the Guatemalan government right after Pearl Harbor. Our purpose was to study the old *Cinchona succirubra* trees which had been planted at El Porvenir during the 1880s, and obtain seed from some of the most vigorous survivors to use as rootstock for grafting at Finca El Naranjo and elsewhere in Guatemala. We were successful in this mission.

As was his custom, Wilson Popenoe had taken his soil auger along with him. Upon examining the soils at El Porvenir, he was most favorably impressed, and he pointed out the volcanic soils on the southwest coast of Guatemala were far more friable and fertile than the limestone soils of Cobán and the Alta Verapaz in the central part of the country.

At the manager's comfortable house, we were given a sumptuous luncheon, served by a barefoot Guatemalan Indian in a spotless white uniform who spoke German.

The cinchona project in Guatemala turned out to be moderately successful. It was an additional crop to aid in the diversification of agricultural production in the country, whose economy at the time was almost entirely dependent on coffee and bananas. Many years later, Merck's main cinchona plantations in Guatemala were purchased by Dutch interests. Merck & Co., Inc. is no longer in the quinine business. There is a moderate demand for quinine at the present time, although synthetics such as chloroquine, primaquine, mefloquine, and pyrimethamine have largely replaced it in malaria therapy. Certain types of malaria, however, are resistant to synthetics, so in some cases quinine is still the drug of choice. Cinchona bark is still being produced in Guatemala at several of the fincas we visited in 1940. It is exported mainly to Germany and Holland.

Chapter XII

Zamorano, Honduras, 1941–1957: Escuela Agrícola Panamericana

In Honduras, in the picturesque valley of the Río Yeguare, twenty-five miles east of Tegucigalpa, there is an agricultural school that has attracted worldwide attention among educators and agriculturists. Known as Escuela Agrícola Panamericana (Pan American Agricultural School), and also in recent years as "Zamorano," it was founded in 1942 with the authorization and support of the government of Honduras, and financed at the beginning by the United Fruit Company, which is now called Chiquita Brands International.

For many years, Samuel Zemurray, president of United Fruit, lived and worked in Central America. The company had made considerable profits in the region; it was time to balance the books—to do something constructive to help Central Americans. An agricultural school would be a worthy project, a place where young men could receive at no expense to themselves sound agricultural training which would enable them to assist in the development of their countries. Zemurray asked Wilson Popenoe, whose judgment he respected, to select the site for the new school and become its first director. Popenoe agreed on the condition that sufficient funds would be made available by the United Fruit Company to make sure the project would succeed. Zemurray assured him it would be a first-class, all-out effort.

For several weeks in 1941 Popenoe roamed the Central American highlands, and finally selected a thirty-five-hundred-acre property called Zamorano, which was purchased from the government of Honduras for the sum of thirty-five thousand dollars. Zamorano turned out to be an excellent choice. In climate and soil it was representative of large areas in Central America where small farmers owned land; five months of rainfall were followed by seven months of dry weather; the soils were typical—not too fertile and not too poor. Its elevation of about twenty-six hundred to fifty-five hundred feet was pleasant for living and ideal for crops; warm enough for tropical lowland crops

Samuel Zemurray, 1877–1961. In 1941, Zemurray, president of the United Fruit Company, decided to establish an agricultural school, Escuela Agrícola Panamericana, at Zamorano, Honduras. Wilson Popenoe was the school's founding director. (*Courtesy of Escuela Agrícola Panamericana.*)

at the lower elevations, yet cool enough for Temperate Zone plantings such as potatoes, peaches, and plums in the higher areas. An adequate supply of pure water for domestic use was available from springs in the upper part of the property. A mountain stream provided some hydroelectric power, as well as enough water to irrigate about two hundred acres. This was important, since the students could learn

Wilson Popenoe in the first office of Escuela Agrícola Panamericana at Tegucigalpa, Honduras, 1942. (*Courtesy of Nancy Popenoe Guillou.*)

about irrigation during the dry season. Zamorano was far enough away from the capital city of Tegucigalpa to provide the quiet, normal life of the farmer, without the distractions and temptations of a city. Tegucigalpa in those days was about an hour and a half away by car over a fairly rough mountain road, which has subsequently been improved. There was ample land in the valley to supply the school with dairy products, vegetables, and other staples; in addition, there was a large pine forest on the property where students could study forestry, learn how to conserve and use timber, and carry out reforestation.

The United Fruit Company was a benevolent founder of Escuela Agrícola Panamericana: to begin with, in 1942 the company donated three million dollars to get the school started; then, over a period of twenty years, it contributed another three million dollars to cover inevitable deficits and pay for capital improvements. Furthermore, the company made it a policy not to hire any graduate of Zamorano in order to emphasize the fact that the school was not going to be used as a training ground to mold recruits for United Fruit's own money-making banana enterprises but for the benefit of Central American agriculture in general.

At first the title "Granja Centroamericana" (Central American Farm) was considered for the school. After much discussion Escuela Agrícola

Valley of Río Yeguare, Zamorano, Honduras, 1945. The first buildings of Escuela Agrícola Panamericana can be seen in the center of the photograph. Hugh Popenoe is in the foreground. (*Courtesy of Hugh Popenoe.*)

Panamericana was chosen since it was planned to include students from many Latin American countries—not just Central America—and emphasize the spirit of pan-Americanism.

From the beginning, the motto of Escuela Agrícola Panamericana has been "aprender haciendo" (learn-by-doing), supplemented by the dictum "labor omnia vincit" (work conquers all). The students were to spend their mornings in the field "getting their hands dirty," and their afternoons in the classroom—a balanced education with emphasis on practical work, not theory. Wilson Popenoe stressed the practical side of education—the learn-by-doing concept. He asserted, "The acquisition of practical skills is the major objective. A student

Escuela Agrícola Panamericana, Zamorano, Honduras, circa 1947. The administration building, Zemurray Hall, can be seen in the foreground. (*Courtesy of Columbus Memorial Library, Washington, D.C.*)

who has learned to milk a cow, to make good butter and cheese, to feed pigs properly, to run a tractor, to protect his soil from erosion by planting on contours, to graft and prune a fruit tree, to grow potatoes, will never forget how to do these things as long as he lives." Then Popenoe pointed out that the student who "crams" for a theoretical examination may forget what he has memorized in a few days.

Several students years after their graduation wrote Popenoe to this effect: "I have a good job now and I am not afraid of it, because at the school I learned to do with my own hands things I am required to supervise here."

Ground was broken and construction started in late 1941. The first four dormitories were named for famous Central American presidents: Morazán of Honduras; Barrios of Guatemala; Mora of Costa Rica; and Delgado of El Salvador. Many young Central Americans were keen to attend Zamorano. One appeared even before the first house had been finished and asked to fill out an application form. He was told to go home and wait: he would be advised in due course if he had been accepted. However, the boy wanted so much to attend the

The motto of Escuela Agrícola Panamericana is "learn by doing." This student is working on an irrigation ditch in a citrus nursery. (*Courtesy of Hunt Institute for Botanical Documentation, Carnegie Mellon University, Pittsburgh, Pa.*)

school, he refused to leave the premises. He was given a temporary construction job until the school opened, and eventually became an outstanding member of the first graduating class.

In September 1943 the Escuela Agrícola Panamericana was opened for a year of tryout and experimentation. The first student body consisted of seventy-three young men from Mexico and Central America.

The students at Escuela Agrícola Panamericana spend their mornings in the field, their afternoons in the classroom. (*Courtesy of Escuela Agrícola Panamericana.*)

Horticulture included dirt-farming experience in planting seeds, rhizomes, and cuttings as well as training in grafting valuable varieties of fruit trees. A year earlier, Popenoe had written to his erstwhile chief at the U.S.D.A., David Fairchild, about the forthcoming program at Zamorano:

> Yes, we must make our boys learn to grow plants, and to know plants. It will be hard on some of them. But I think I'll start them out like this: every boy, the first month he is at the school, has to start a small nursery, planting seeds of sour orange, mango, and avocado (perhaps others); he has to carry these plants through the nursery stage and graft them, then train the grafted plants. By the time he has finished his 3 years at the school, he will have at least 500 fine grafted fruit trees which he will be free to take home with him, give away, or sell.

On Columbus Day, 1944, the Escuela Agrícola Panamericana was formally opened with about 140 students. In the United States, with so many diverse colleges and universities, another agricultural school inauguration would not have been especially noteworthy. But it was a memorable occasion for Central America, a part of the continent that at the time was lacking in facilities for agricultural education. Most educational programs in universities of that region concentrated on law, medicine, and the classics, even though agriculture represented

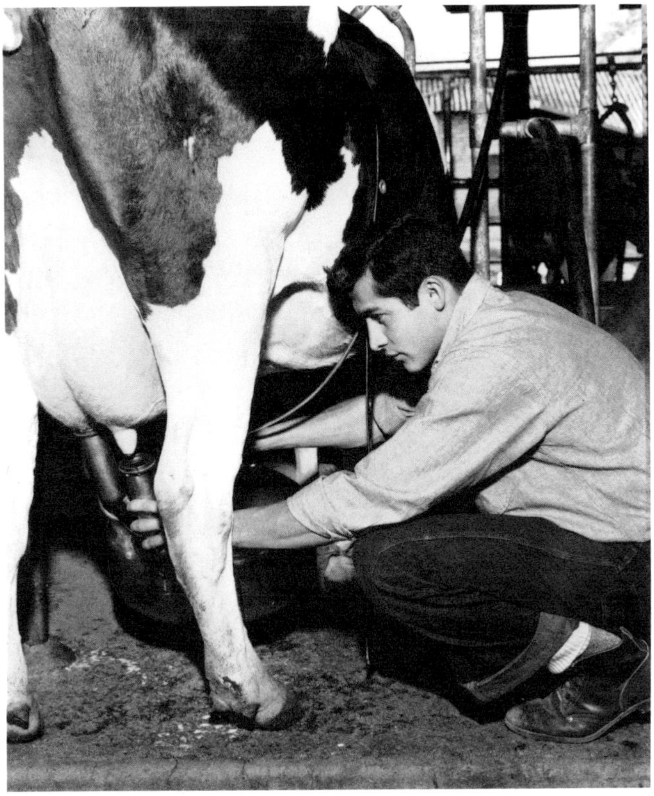

Taking care of dairy herds is part of the program at Escuela Agrícola Panamericana. (*Courtesy of Escuela Agrícola Panamericana.*)

at least 80 percent of gross national production and provided employment for more than half the population.

In the first year, during the mornings, the students worked in the horticultural department, where they acquired the skills necessary for the production of vegetables and the care of fruit trees. During the second year, in the mornings, the emphasis was on field crops: the planting, care, and harvesting of such crops as corn, beans, sugar cane, rice, and sorghums. The mornings of the third year were devoted to animal husbandry, to learn to take care of dairy herds, to pasteurize milk and make butter and cheese, to care for hogs and prepare ham and bacon, to butcher steers, to cut meat properly, to raise chickens, and to treat sick animals. During the afternoons, classroom work was devoted to mathematics, biology, chemistry, English, field and horticultural crops, farm management, and the use and care of farm machinery. Spanish has always been the accredited language of the

Bee keeping is part of agricultural training at Zamorano. (*Courtesy of Columbus Memorial Library, Washington, D.C.*)

school. The school paid the basic expenses of the students, including food and housing, dental and medical services, clothing, books, and other essential supplies.

At Zamorano, the main building where the students congregated was called Zemurray Hall; it housed the administration office, the library, laboratories, classrooms, and the assembly hall, which also served as a movie theater on Friday nights and as a church on Sundays. The architectural style of the campus was colonial Spanish; the buildings were constructed of soft volcanic stone quarried on the property, with stuccoed walls, red tile roofs, and handsome wooden doors, hand carved by skilled Honduran carpenters. Baked floor tile was used in most houses; construction details were patterned after well-proportioned, traditional buildings in Honduras and Guatemala.

Part of the dawn-to-dusk schedule at Zamorano includes work in the chemical laboratory. (*Courtesy of Hunt Institute for Botanical Documentation, Carnegie Mellon University, Pittsburgh, Pa.*)

At first the students were mainly farm boys between the ages of sixteen and twenty-one, from families of limited means. Entrance requirements were determined by the primary school education available at the pupil's home community. In rural areas of Central America this amounted to roughly four or five grades of public school in the Unites States. The original prescribed course of study was for three years, totaling twenty-seven months. The school year started in June and continued through the month of March.

Wilson Popenoe believed in hard work and learning by doing—not in degrees as a status symbol. In fact in 1942 he did not want to award any degree at all to graduates of Zamorano; he feared students might then rest on their laurels and end up in desk jobs. To students who wanted a degree he quoted the Scriptures: "By their works ye shall know them"; and he reminded the student body that the Imperial College of Tropical Agriculture in Trinidad, far and away the outstanding agricultural school of tropical America at the time, gave no degree, just a certificate to the effect that the student had satisfactorily completed the course of study. A compromise was reached at Zamorano with the awarding to graduates of a certificate entitled "Agrónomo," when they completed the three-year course.

Although most students were poor farm boys, some did come from more affluent families. Popenoe emphasized an important point: "It is probably safe to say that no feature of our program has done more to develop a genuine spirit of democracy than the manual labor of boys from wealthy and influential families working side by side with boys from the humblest homes."

Wilson Popenoe's method of teaching horticulture was simple and direct. "Here is the classic text on grafting," he would say as he held up a book for the class to see. "It shows 125 different ways to graft, but all you need to learn is five—and now we will go out and learn them." Taking the students out to the nursery he would say: "Watch me. This is how you slice off the buds," as he sliced off one after another with an effortless rhythm. Some students would have problems as they ineffectively hacked and gouged the scions and rootstock. "Don't give up. Don't be discouraged," he would tell them. "After you have sliced a thousand buds, you will find you can do it smoothly." Popenoe had a fast way of talking, so irreverent students used to call him "Dr. Rápido."

The lack of textbooks and technical journals in Spanish made English a necessary subject—which would also come in handy in case graduates should continue their studies in the United States.

One of Wilson Popenoe's favorite stories was about an English class in which he asked the students what they would do if cast up, all alone, on an uninhabited island of the Caribbean, and they could take with them just one crop plant, one fruit tree, and one animal. What would they be? Julian Meléndez was the first student called on. He replied as follows: "Well, I will take the corn plant because corn is the basis of our Central American diet. And for the fruit tree, I will take an orange because I like oranges and the fruit will supply me with vitamins. And since I will be on an island, and can catch all the fish I want, I won't need any meat, so for the animal I will take a woman, so I won't be lonesome."

On another occasion, Wilson gave an examination where the following question was asked: if you want to grow pineapples, what variety would you choose, what climatic and soil conditions would you look for, and what would be your production practices? One student answered that he didn't want to grow pineapples. Wilson gave him an A!

On March 19, 1952, David Fairchild, in his role as president emer-

Students learning plant propagation as they graft selected citrus scions onto rootstock in the Zamorano nurseries. (*Photograph by Teodoro Albornoz.*)

itus of Fairchild Tropical Garden, offered Wilson Popenoe a lucrative job in Coconut Grove, Florida. Popenoe replied as follows on March 29, 1952, showing his devotion to Escuela Agrícola Panamericana.

Dear Chief:

You can well imagine how your letter of 19 March pulled at the heartstrings both of Helen and myself. Our devotion to the Fairchilds, which has never lessened but on the contrary has increased throughout the years, makes it hard not to go to you immediately when you think you need me.

Both of us would like the job. I am not wedded to California nor do we ever expect to live there permanently. For many years I have said that if I ever retired and could afford it I would like to live in Coconut Grove, primarily because I could really grow tropical plants there, and secondly because there are so many fine people in that colony.

If I had been smart enough to bring somebody along here at the start to take over the job at any time, I would seriously consider pulling out right now and moving to Coconut Grove. But the unfortunate fact is that I have not been able to find anyone yet who can handle this job the way I wish to see it handled. Perhaps my way is not the best. I have no illusions on that point, but at the same time I feel I have a real obligation to carry on here as long as I am able and as long as the Company is satisfied.

Only two days ago one of our men was up here from the coast and I was telling him that we simply must get somebody in line to take over this job. He replied that he had recently overheard Mr. Zemurray dis-

Wilson Popenoe, the educator, with his faithful soil auger. (*Courtesy of Hunt Institute for Botanical Documentation, Carnegie Mellon University, Pittsburgh, Pa.*)

cussing the matter and that he had said: "Oh, don't worry about the School, Pop is good for a long time yet." Mr. Zemurray and the other high officials of the Company have asked me time and again to stay by this job just as long as I am physically able to do so.

I feel an obligation to the Company where I have been well treated. I also feel an obligation to these boys in whom I am so deeply interested and who, I feel, are really being helped to make better farmers and better citizens of themselves. The need is so great; the field is so important from more than one angle. I have put my hand to the plow and I just simply cannot turn back.

I have to spend at least five years more here, at which time I will be 65, and I would like to spend ten years more here if I live and keep my health. By that time I won't be of any use to the Fairchild Tropical Garden or anything else.

I hate to say "no." I hate to miss that wonderful opportunity to work all day long with tropical plants instead of having to handle a thousand and one petty administrative details. It is not a matter of salary. It is

The coeducational program at Zamorano began in 1981. Today 10 percent of the students are women, and the percentage is increasing. (*Courtesy of Escuela Agrícola Panamericana.*)

solely a matter of having started something here which I feel, rightly or wrongly, still needs me.

And finally, both Helen and I hate to miss the opportunity of living close by the Fairchilds once again.

All in all, the decision we have to make is a tough one.

Affectionately yours,
Wilson

Reforestation work is an important part of the school curriculum at Zamorano. (*Courtesy of Escuela Agrícola Panamericana.*)

In August 1954, Wilson received a sad telegram from his "godmother," Marian Bell Fairchild: "Your beloved chief has set out on his last trip." David Fairchild had died at his home, "The Kampong," in Coconut Grove, Florida, at the age of eighty-five.

Wilson Popenoe retired as director of Escuela Agrícola Panamericana in 1957 at the age of sixty-five, following sixteen years of admirable leadership. The school library has been named Biblioteca

Vice-President and Mrs. Nixon visited Zamorano in 1955. (*Courtesy of Escuela Agrícola Panamericana.*)

Wilson Popenoe in his honor and memory. Although he moved to the Casa Popenoe in Antigua, Guatemala, he remained in close contact with the school at Zamorano for many years as a trustee.

Since 1957 the school has enjoyed steady growth, as it continues to stress the unique characteristics that set it apart from other agricultural institutions in Latin America: namely, the emphasis on hands-on training and especially on "learning by doing." Tropical farming is constantly and far more severely threatened by hurricanes, floods, erosion, plant diseases, and pests than agriculture in temperate zones. Sound knowledge of the most modern agricultural techniques is essential for young farmers to succeed in the tropics. Therefore, a fourth year of specialized courses was added at Zamorano in 1987 to enable some students to obtain the degree of "Ingeniero Agrónomo" (agricultural engineer). Technical work and individual research are emphasized in the fourth year. A thesis is required in the "Ingeniero Agrónomo" program.

John G. Smith and Simón E. Malo. During the decade of the 1980s, Smith was chairman of the board of trustees of Escuela Agrícola Panamericana; Malo was the school's director. (*Courtesy of Escuela Agrícola Panamericana.*)

A number of able educators have carried on the work as directors of the school at Zamorano since Wilson Popenoe's departure, including William C. Paddock, Albert S. Muller, Robert P. Armour, Kermit H. Adams, Joseph S. Courand, and Simón E. Malo. Dr. Malo has performed most effectively as the school's director since 1979; during this period the size of the student body has tripled to over six hundred, of whom about 60 percent receive scholarship assistance.

Zamorano now has a number of water buffaloes (*Bubalus bubalus*) of Asian origin. These are versatile farm animals whose importance is growing in Central America. They can thrive on poorer quality forage than cattle, and buffalo meat is cheaper to produce. (*Photograph by Teodoro Albornoz.*)

Since 1981 Zamorano has actively recruited female students. An effort is being made to increase the proportion of young women to young men from the current 10 percent to 25 percent. Zamorano has a high-quality faculty with a concentration of PhDs far greater than most agricultural schools in Latin America. The student-to-professor ratio has traditionally been about ten to one. Most of the faculty live on the campus and are continuously available for advice and consultation. The faculty, the administration, and the student body are apolitical.

The policies of Escuela Agrícola Panamericana are governed by an international board of trustees, whose members, drawn from the private business and academic sectors of North and Latin America, meet several times a year. Since 1942 some distinguished individuals have served as chairmen of the board of trustees, including the late Samuel Zemurray; his daughter, Mrs. Doris Zemurray Stone; the late J. George Harrar, former president of the Rockefeller Foundation; the late Francisco de Sola, a prominent industrialist and agriculturist from El Salvador; J. Wayne Reitz, former president of the University of Florida; Hugh Popenoe, Wilson Popenoe's son, director of the Center for Tropical Agriculture, University of Florida; John G. Smith, a forward-

Escuela Agrícola Panamericana has graduates from twenty-two countries throughout the world. The students for the most part come from Central America, Ecuador, Colombia, and Bolivia. (*Courtesy of Escuela Agrícola Panamericana.*)

looking coffee grower from Guatemala; and Duncan H. Cameron, senior partner of a law firm in Washington, D.C.

Escuela Agrícola Panamericana is a private institution with a comparatively large endowment and a broad base of donors. Important benefactors and cooperators include some major governmental and international organizations: the United States government through USAID/ASHA (American Schools and Hospitals Abroad); USAID missions to Honduras, Guatemala, and Panama; and USAID/ROCAP, Guatemala (Regional Office for Central America and Panama). The Federal Republic of Germany has contributed significantly through many scholarships. Generous support has also been received from the government of Honduras.

Major foundations and corporate donors for many years, apart from Chiquita Brands International (United Fruit Company), include the Rockefeller Foundation, the W. K. Kellogg Foundation, Cargill Incorporated, the Bank of Boston, Exxon Corporation, Texaco Incorporated, the Gillette Company, and Deere and Company, to name a few. Over the years, major individual supporters of the school have included the late Thomas Jefferson Coolidge; his daughter, Dr. Catherine Coolidge Lastavica; the late Charles L. Stillman; Thomas D. Cabot; Mrs. Doris Zemurray Stone; her son, Dr. Samuel Stone; John R. Kimberly; George P. Gardner; Adolfo Midence Soto; and Marcel Laniado de Wind. Rafael Aguilar Paz donated a total of 4,280 acres of land known as "Rapaco" to the school in 1975 and 1977.

The activities of Escuela Agrícola Panamericana have expanded in recent years. Courses in computer science and aquaculture (fish farming) have been added. An International Seed and Grain Science Center has been established at the school. Zamorano produces one thousand tons of seeds annually, including over 50 percent of the certified corn seed of Honduras. Sizable numbers of beef cattle, dairy cattle, swine, goats, sheep, water buffaloes, and poultry are maintained for research and training. Basic outreach programs aimed at improving the agricultural productivity of Central America are given, including short courses in agricultural extension work. Food production is vital in the region, where the annual population growth rate is now 3.2 percent. Fewer farmers have to produce more and more food for the growing population in urban areas. More qualified educators are needed to teach and demonstrate improved farming methods. Zamorano plays an important role in this educational program. Several other agricultural schools in Latin America have been modeled after Escuela

Agrícola Panamericana, including two in Costa Rica and one in El Salvador.

Zamorano's rural setting consists today of a large property of about fifteen thousand acres, including forest reserves. Substantial farm and forest production, including reforestation, is carried out with active student participation. The size of the property of the school has doubled during the past ten years.

There are now over three thousand graduates of Escuela Agrícola Panamericana distributed throughout more than twenty countries. Many have had successful careers: twenty-five have become ministers or vice-ministers of agriculture in the governments of their respective Latin American republics. Other prominent graduates include bankers, professors, scientists, and university presidents.

The Escuela Agrícola Panamericana at Zamorano, Honduras, expects its students to have a strong obligation to high academic achievement. It demands unceasing dedication and performance in the field, classroom, and laboratory. There is rigorous discipline, as well as a commitment to building character and maturity through honesty, self-reliance, and hard work. It has become a model agricultural school for the tropical world. Wilson Popenoe would be pleased with the progress that has been made in fifty years by those who have "learned by doing."

Chapter XIII

Retirement Years, 1958–1975

After Wilson retired as director of Escuela Agrícola Panamericana in May 1957, he became director emeritus and consultant to the board of trustees of the school. His retirement started off on an upbeat: he received highly congratulatory letters from both President Eisenhower and Vice-President Nixon. Eisenhower's letter appears on the next page.

On the way back to Guatemala, Wilson stopped at El Salvador where, at the request of the government, he made a study of the possibilities for tropical fruit culture in that country. He started a program which led ultimately to the development of several plantings of tropical and subtropical fruits at various elevations in El Salvador. He also submitted a design for the National Agricultural School of El Salvador, based on his broad experience at Escuela Agrícola Panamericana.

Wilson and Helen, his second wife, whom he had married in 1939, then returned to the Casa Popenoe in Antigua, Guatemala, for a few months of well-deserved rest. Helen had been a devoted wife and a great help to Wilson in many ways. She had constantly given him support during fifteen years at Zamorano in the sometimes stressful role of wife of the school's founding director; she was a gracious and untiring hostess, welcoming visiting dignitaries to Escuela Agrícola Panamericana. When Vice-President and Mrs. Nixon visited Honduras in 1955, the Popenoes were obliged to entertain hundreds of visitors at Zamorano. Now, in his retirement, she was well qualified to assist Wilson in the continuing restoration of the Casa Popenoe as they gradually filled it with carefully selected antique furniture, paintings of early Guatemalan dignitaries, and early Mayan pottery, silverware, and artifacts.

Before coming to Guatemala for the first time in the summer of 1938, Helen Barsaloux, at the age of thirty-four, had been an art buyer for several years for the Chicago Museum of Art. She spoke Spanish well since she had studied for two years at a university in Madrid.

THE WHITE HOUSE

WASHINGTON

May 23, 1957

Dear Dr. Popenoe:

From many sources I have heard of the fine work you have done as Director of the Escuela Agricola Panamericana, and now I understand that, on the occasion of your retirement, you are to become Director Emeritus. I hope you will permit me to join with the many distinguished Americans who salute you not only for your outstanding contribution to hemispheric understanding through your work at the School, but for your many achievements in furthering sound agricultural practices throughout the world.

The Escuela Agricola Panamericana, of course, excites my admiration because it so practically demonstrates ways in which the technical knowledge we have gained can be employed to increase the productivity and raise the living standards of our good neighbors to the south.

With best wishes,

Sincerely,

Dr. Wilson Popenoe
Director
Escuela Agricola Panamericana
Zamorano, Honduras

(*Courtesy of Hugh Popenoe.*)

Wilson Popenoe with his second wife, the former Helen Barsaloux, whom he married in 1939. (*Courtesy of Hunt Institute for Botanical Documentation, Carnegie Mellon University, Pittsburgh, Pa.*)

Having read Louis Adamic's *The House in Antigua,* Helen and a lady friend from Chicago decided to visit Guatemala. They called at the Casa Popenoe in Antigua, where Wilson himself showed them around. During the next ten days Wilson and Helen spent a lot of time together.

A few months later, in early January 1939, they were married, she for the first time.

Wilson received a brief letter of congratulations from David Fairchild, dated January 14, 1939: "So you have married again. I am glad, very glad indeed, for the thought of you wandering about like a lonesome ghost has always haunted me. . . . Give her our love and assure her from me that I think she has won a prize."

Early in 1958, accompanied by Helen, Wilson transferred his activities for a year to the Mediterranean coast of Spain, where he devoted himself to the introduction of the world's best commercial avocado varieties for propagation in Spanish nurseries. They stayed in the small Moorish town of Almuñecar, a Mediterranean seaport in southern Spain in the province of Granada, where they enjoyed the countryside. Wilson wrote Mrs. Fairchild:

> What a place, Almuñecar. As long as we live we shall see ourselves sitting on the wall in front of our little apartment, looking out over the valley with its cherimoya orchards and fields of sugar cane, to the red hillsides covered with almond and olive trees and the little white houses—Oh those lovely little white Andalusian houses! If I were a young architect I would spend a year making measured drawings of them and taking photographs, and I would go back and literally fill California full of them.

Wilson succeeded in establishing twenty avocado varieties in southern Spain, in addition to some macadamia plants, Mexican limes, and mangos from Israel. He visited the Canaries, where he introduced the Haden mango to Tenerife. Then he made a trip to Israel where he assisted the hard-working Jewish banana growers with their drainage problems—with which he had been so familiar in Central America in his years with the United Fruit Company.

Recognized as a world authority on tropical and subtropical fruits, Wilson wrote several articles in 1958 from Spain for the *Encyclopedia Britannica,* in which he covered the guava, cherimoya, mangosteen, papaya, avocado, and banana.

In early 1959, after they had returned to Antigua, Guatemala, Wilson received a lucrative offer to take over the direction of the National Agricultural School in El Salvador for a period of two years. He turned it down for two reasons: (1) he didn't like the local politics at that time in El Salvador; and (2) he felt at sixty-seven he was too old for the job. He quoted an old Jamaican proverb: "Old jackass no b'long in hoss race."

Wilson now turned his attention to the development of temperate-latitude fruits in the highlands of Guatemala, between thirty-five hun-

The Guatemalan artist Humberto Garavito painting a portrait of Wilson Popenoe at Casa Popenoe, about 1950. (*Courtesy of Nancy Popenoe Guillou.*)

dred and seventy-five hundred feet above sea level. He was enthusiastic about the future production of apples, plums, and peaches. He pointed out in a booklet entitled "Temperate Zone Fruits in the Central American Highlands" which he wrote in 1962, in cooperation with Jorge Benítez, that the introduction of European fruits to Spanish America was started in the colonial period. Hernán Cortés wrote a letter to Emperor Carlos I in 1524: "I have already made known to your Imperial Majesty the necessity of bringing to these lands plants of all kinds . . . and I repeat the request, hoping that orders can be given to the Casa de Contratación in Sevilla to the effect that every ship must bring plants, and that no ship shall be allowed to depart without them."

It is believed the first fruit plants brought by the Spanish were olives and grapes, which did not do well in Central America but flourished in Chile and California; then peaches, which did very well indeed, and have become the most widely distributed and most popular of the Temperate Zone fruits. It seems probable that apples, pears, and plums came somewhat later. In recent years, improved varieties of apples, plums, and peaches have been grown successfully in the Gua-

Garavito's portrait of Wilson Popenoe with avocado. (*Courtesy of Nancy Popenoe Guillou.*)

temalan highlands; there have been problems with pears and grapes. Some progress has been made in growing plums, strawberries, raspberries, youngberries, and the Andes berry (*Rubus glaucus*).

In conclusion, Wilson observed that future progress in growing fruits in the highlands of Guatemala and the other Central American countries now depended largely on the plant breeders—who had plenty of material to work with. Considerable information was available about insects, plant diseases, slope exposure, chilling requirements, length and dates of rest period, diurnal temperature changes, photoperiodism (the reaction of the plants to variation in day length by flowering or not flowering), cross-pollination, and other factors such as transportation and marketing.

In the late 1950s Wilson persuaded the Guatemalan government to establish experiment stations in the highlands to multiply these Temperate Zone fruits and distribute the plants gratis to the Indians. For several years he spent a lot of time at these experiment stations

Paul Popenoe (left) and his brother Wilson, Altadena, California, 1945. Paul Popenoe (1888–1979) was an author, social biologist, newspaper columnist, and expert on marital relations. A pioneer in family counseling, he was the author of many articles and books, including *Modern Marriage* (1925); *Marriage, Before and After* (1943); and *Marriage Is What You Make It* (1950). (*Courtesy of Nancy Popenoe Guillou.*)

teaching the personnel how to bud and graft selected varieties of the different fruits.

Wilson loved to banter with the market women. On one occasion in Antigua, he wanted to buy one hundred avocados to use as rootstock for grafting in a nursery. He asked a venerable Indian who had been sitting around all day trying to sell her wares, how much the avocados cost. She answered they cost one cent each. So he said, "All right, I'll take one hundred, here's one dollar." "No," replied the Indian, "it will cost you one dollar and twenty-five cents. If you can buy one hundred, you must be rich and can afford to pay one dollar and twenty-five cents." So Wilson paid one dollar and twenty-five cents with a smile. He had a delightful, enthusiastic way of telling jokes in fluent Spanish with a trace of a gringo accent. It is hard to represent his humor through the written word. When he told a story, the people he described became alive. With all due modesty, he said he was in demand as a public speaker for three reasons: (1) his talks were short; (2) he told jokes; and (3) he made no charge.

Here is another of Wilson's stories. During the 1930s there was a famous German priest at Chichicastenango in the highlands of Guatemala named Ildefonso Rossbach—who did a great deal to help the Indians, spiritually and physically. Late one afternoon in 1940, the padre, standing on the steps of the church, was addressing his parishioners. Blessing them in a monotonous Latin prayer, he noticed Wilson among the dozens of Indians in his audience. He continued in singsong Latin: "In nomine Patris et Filii et Spiritus Sancti, Friend Wilson go into my study—there's a bottle of Johnny Walker Scotch on the sideboard—pour yourself a drink. Amen." Lifting up their bowed heads, the Indians cried out "Amen."

Wilson Popenoe didn't mind being called a "gringo." For him the epithet was an asset, not an insult, as he explained to his friend Nixon Smiley, who for years was farm and garden editor of the *Miami Herald*:

> Everybody in the world seems to have an unflattering name to call the other fellow. Southerners call northerners "Damn Yankees" and northerners call southerners "rebels." And there are local names, like "tarheel" for Carolinians and "crackers" for Georgians and Floridians. When I introduce myself to Latins I may say: "I'm Wilson Popenoe. I'm a gringo." This invariably brings a laugh. The Latins laugh and I laugh. When people laugh they become more at ease. I call myself a gringo before they have a chance to do so. They know that I have no feeling of superiority—that I want to be one of them.

Wilson went on to explain the origin of the word gringo. According to him, gringo comes from griego, the Spanish name for Greek. In early Spain, so many Mediterranean travelers were Greeks that the Spaniards began to call all foreigners griegos. In the course of time, griego became gringo. He didn't agree with the more common explanation that the name originated from a song which was sung by General Zachary Taylor's troops in Mexico in 1845 called "Green Grow the Rushes, Oh." Wilson said this can't be the true origin, since the word gringo had already been in the Spanish language for several centuries, long before 1845. True or false, Wilson was an exceptional gringo ambassador of good will in Latin America.

In September 1959 Wilson visited Venezuela at the request of Sr.

Eugenio Mendoza. He described this trip in a letter to Mrs. David Fairchild dated October 15, 1959:

> I went to Venezuela on a very nice job. It was the sort of job that would have warmed the cockles of David Fairchild's heart. A very fine, and very wealthy man, Eugenio Mendoza, has a philanthropic project which is the improvement of fruit cultivation in Venezuela. . . . Don Eugenio invited me down there to study the possibilities and plan a fruit improvement program and I told him I would come, and just for my expenses, if he would agree to carry out the program after we had talked it over and he was satisfied with it. Things went well, and I tell you one thing, I am glad I did not ask for any honorarium, because the traveling expenses were astronomical. Not my fault either; in fact it became a joke among the brethren that I moved out of the Hotel Maracay because the prices were so high they killed my appetite.

Few horticulturists and educators have been honored as much as Wilson Popenoe. In Antigua he had a small trunk filled with medals, decorations, and miscellaneous awards that had been bestowed on him by numerous governments and educational institutions. He received the Orden de Rubén Darío from Nicaragua, the Orden de Francisco Morazán from Honduras, and the Orden del Quetzal from Guatemala. He was on a first-name basis with all the Central American presidents in the 1930s and 1940s. When the author stayed at the Casa Popenoe in 1940, there were several Mexican, black-bellied, whistling tree ducks wandering around the main patio: one was named "Tacho" for President Anastasio Somoza, Senior, of Nicaragua; another "Tiburcio" for President Tiburcio Carias of Honduras.

The following list covers some of his other well deserved honors: in 1923, the Chilean Order of Merit for furthering the interchange of useful plants between the United States and Chile; in that same year, the Société Nationale d'Acclimatation de France awarded him a Silver Medal for Plant Introduction work; Ecuador conferred on him the Order of Merit; Panama the Order of Vasco Nuñez de Balboa; Cuba the Order of Agricultural and Commercial Merit; the University of San Carlos in Guatemala named him an honorary professor. The city of Antigua, Guatemala, gave him a gold medal in recognition of his work in agriculture and for restoration of the Casa Popenoe—which became such a popular attraction for tourists.

In the United States he was granted the honorary degree of Doctor of Science in 1947 by Pomona College, and in 1960 his third honorary doctorate, by the University of Florida. In the year 1950 he received two other important awards: the George Robert White Medal of Hon-

In June 1947, Wilson Popenoe's daughter Nancy graduated from Pomona College, while he received an honorary degree. (*Courtesy of Paul Popenoe.*)

or from the Massachusetts Horticultural Society, for distinguished services to horticulture in both the United States and Latin America; and the Frank N. Meyer Medal—named in memory of an outstanding plant explorer of the U.S.D.A. who died in China in 1918—for plant exploration and plant introduction, from the American Genetic Society. In 1960 he was given the Wilder Medal by the American Pomological Society for "leadership in Tropical and Sub-tropical Fruits." There were many other decorations and citations in addition to the honors mentioned above; one was the Award of Honor from the California Avocado Society for his developmental work in the avocado industry in California.

In 1960, Wilson's second son, Hugh, received his PhD from the University of Florida. Hugh, a bachelor, is the director of the Center for Tropical Agriculture at the University of Florida and is president of the American Water Buffalo Association. During the 1950s Wilson's three daughters were all married. Nancy became Mrs. Robert Guillou, Marion was wedded to Richard Hatch, and Sally married Edmund Hally, Jr. By the mid-1960s Wilson had acquired nine grand-

Wilson Popenoe; his second wife, Helen; Sally; Marion; and Hugh; 1944. (*Courtesy of Hunt Institute for Botanical Documentation, Carnegie Mellon University, Pittsburgh, Pa.*)

children. Unfortunately, Guillou died in a helicopter accident in Utah in 1965; Hatch was also killed in a helicopter, during combat in 1968 in Vietnam. Marion now has a PhD and is head of the department of archaeology at the University of the Valley of Guatemala. Regrettably, Wilson's first son, Peter, was retarded and completely incapacitated. For many years he was in a private institution near Los Angeles where he passed away in 1971 at the age of forty-seven.

Wilson's younger brother, Herbert Franklin Popenoe, an educational psychologist with the Los Angeles school system, died in 1975 at the age of seventy-six. Wilson always maintained that shy, mild-mannered Herbert, who had an unusually high IQ, was the most intelligent of the three Popenoe brothers.

In March 1961 Helen Popenoe died suddenly in Antigua at the age of fifty-seven, following a massive stroke. This was the second time Wilson had lost a beloved wife unexpectedly. He was emotionally and physically drained. Three months later he was hospitalized in Guatemala for gastrohepatic trouble. After he got out of the hospital, he went to Florida to live for a year with his son Hugh at Gainesville. While in Florida he was injured in an automobile accident which further debilitated him. In due course, his health was sufficiently restored to enable him to return to Guatemala. Back in the familiar Latin atmosphere of the Casa Popenoe in Antigua, his spirits rose and he recovered his sense of humor as well as his enthusiasm for life. He kept himself busy writing hundreds of letters and traveling about the highlands of Guatemala, observing how the Indians were growing and harvesting their corn—the main staple of their diet and an indispensable part of their existence.

Wilson had been interested in corn for many years. In June 1937, he and J. H. Kempton of the U.S.D.A. wrote an article entitled "Teocinte in Guatemala" for the Carnegie Institution of Washington and the Bureau of Plant Industry of the U.S.D.A. They postulated that the tall, cornlike fodder grass *Zea mays* subsp. *mexicana,* commonly known in Guatemala and Mexico as *teocinte,* may be considered as directly involved in the creation of maize, or corn, *Zea mays* subsp. *mays.* Furthermore, teocinte probably originated in the Mayan highlands of western Guatemala rather than in the Aztec tablelands of Mexico. (Some plant biologists today consider maize to be a domesticated form of teocinte.)

Henry A. Wallace, the former vice-president of the United States, and Wilson were good friends, and Wallace stayed at the Casa Popenoe in Antigua several times. On one occasion, Wilson asked Wallace in what year he had left the vice-presidency. Wallace replied: "I don't remember, but whenever it was, it wasn't soon enough." Quite apart from politics, Wallace was a brilliant plant geneticist who had founded a highly successful hybrid corn seed company in 1926. Both Wallace and Popenoe wanted to improve the lot of the poor Indian in the highlands of Guatemala where corn production averaged only about ten to fifteen bushels to the acre, mainly on exhausted land that had been under cultivation for two thousand years or more. During 1963 and 1964 they worked together in a joint effort, sometimes referred to as the Wallace-Popenoe corn project. Their objective was to help the Indians double their production of corn through improved seed and the proper application of fertilizer and insecticides. In cooperation with the Ministry of Agriculture of the Guatemalan government, and Francisco de Sola, an industrialist from El Salvador who provided free fertilizer for the trials, some good preliminary results were obtained. Unfortunately, Wallace, who was contributing financial support as well as technical advice for the project, died in 1965; otherwise, more promising varieties of seed corn for the Guatemalan highlands might have been developed at that time.

During his last visit to the Casa Popenoe, Wallace gave Wilson a pair of handsome hiking boots, but they were too large for him. Wilson used to enjoy telling the story—his feet were not large enough to fill vice-presidential shoes.

In 1966 an international symposium on plant introduction was held at the Escuela Agrícola Panamericana in Zamorano, Honduras, with funds provided by the Rockefeller Foundation. Twenty-two scientists gathered together from different parts of the world, not only to present erudite papers at a scientific meeting, but also to honor Wilson Popenoe on the grounds of the school he founded. The event marked a milestone in the history of the school: Director Albert S. Muller presented the one thousandth diploma to a graduate. Wilson was granted the title of "Agrónomo Honorario" (honorary agronomist), while due recognition was given to his outstanding career as an educator and plant explorer. His old friend Knowles Ryerson delivered a fitting eulogy:

> Among the names of plant explorers is that of Wilson Popenoe, in whose honor we are convened here. For more than a dozen years he followed the trails up, down and through valleys, through heat and cold, almost always alone, seeking avocados, citrus and other fruits of value to his country and its tropical possessions. In the absence of David Fairchild, he directed the activities of the Office of Seed and Plant Introduction. He

Wilson Popenoe with his third wife, the former Alice Weiss, Zamorano, Honduras, 1969. (*Courtesy of Nancy Popenoe Guillou.*)

> has left his initials carved deeply on our horticulture with a professional career any plantsman would envy. But, he has done more. Popenoe has been an American Ambassador of Good Will to Latin America. He always left a trail that others would follow more easily because he had been there. . . . He has crowned his career with the organization and the direction of Escuela Agrícola Panamericana, and has aided in its progress as an advisor since his retirement. As valuable as his many plant introductions have been, I venture to say that his imprint and inspiration in the lives of the many students fortunate to have known him will in the years ahead be his greatest contribution, for people—more than plants—count first.

Many years earlier Wilson had discovered some new plants which were named after him, including *Begonia popenoei*; *Dahlia popenovii,* which he found in Guatemala; and *Passiflora popenoei,* an Ecuadorean vine with an edible fruit of delicious fragrance. Wilson had noticed these fruits being sold by market women in Ecuador in 1921.

In his youth Wilson had the wanderlust—the urge to travel. In 1918, after returning to Washington from a lengthy plant-exploration trip to Guatemala, he wrote a thoughtful poem entitled "Wanderlust":

WANDERLUST

I often think that I would freer be
Could I but rid myself of every taint
Of Wanderlust, which coursing through my veins
Doth surge and struggle, and triumphal, paint
In lurid colors scenes that make me yearn
To cross the stormy waters, comforts spurn,
And once again to seek for happiness
On distant shores, and ever striving, press
Onward into the deepest wilderness.

And on and on to go; I know not Why
Nor Whither, but in vain pursuit of nought
Borne forward by an all-consuming fire
Until at last, my costly freedom bought
I turn my weary footsteps once again
Toward Home. Through valley, over plain
I struggle onward, till I reach my goal
And there I rest for aye, a ransomed soul.

To you, who never yet have felt
That Call of Calls, the Wanderlust, I say
Beware! For once within its grasp
No power in Heaven or Earth can help you; Nay
Your fate is sealed, and forward you must go
To quench the awful flames, nor do you know
The Whither nor the Why; Until at length
The demon's grip is loosened, and his strength
No longer holds you helpless. Then once more
You rest in peace upon your native shore.

In 1969 Wilson remarried: his third wife was Alice Weiss, a long-time friend of the Popenoe family in Honduras, whose father had been the Swiss consul in Tegucigalpa. Alice helped him run the Casa Popenoe and was a cordial hostess at his side when visitors called. He spent the last years of his life in quiet retirement in Antigua. From time to time, he received welcome letters from graduates of Escuela Agrícola Panamericana, many of whom had achieved success in their respective fields. These reports were a source of great satisfaction to him.

Wilson Popenoe passed away in his sleep at the Casa Popenoe on June 20, 1975, at the age of eighty-three. He was buried in the cemetery in Antigua, next to his second wife, Helen.

Principal Publications of Wilson Popenoe 1911–1971

1911. The Avocado in Southern California. Pomona College Journal of Economic Botany I (1): 3–24.

1911. The White Sapote. Pomona College Journal of Economic Botany I (2): 83–90.

1911. The Mango in Southern California. Pomona College Journal of Economic Botany I (4): 154–200.

1912. *Feijoa Sellowiana*: Its History, Culture and Varieties. Pomona College Journal of Economic Botany II (1): 217–242.

1912. The Cherimoya in California. Pomona College Journal of Economic Botany II (2): 277–300.

1913. A Basis for the Future Classification of the Mango. Proceedings of the American Pomological Society: 41–50.

1914. Tropical and Subtropical Fruits in California. Journal of the Royal Horticultural Society 39 (2): 330–337.

1914. Fruit Culture in Baghdad. Gardners' Chronicle 1440: 229–230.

1914. The Jaboticaba. Journal of Heredity V (7): 318–326.

1915. The Avocado in Florida and Other Lands. Yearbook of the California Avocado Association: 29–42.

1915. The Mango of Cuba. Proceedings of the American Pomological Society: 1–26.

1915. Plant Breeding in Cuba. Journal of Heredity VI (12): 558–568 (with F. S. Earle).

1916. The Freezing Point Lowering of the Leaf Sap of the Horticultural Types of *Persea americana.* Journal of Agricultural Research VII (6): 261–268 (with J. Arthur Harris).

1916. The Pitanga. Journal of Heredity VII (4): 179–185 (with A. D. Shamel).

1916. Tropical Pomology. University of California Journal of Agriculture III (6): 221–228.

1917. The Navel Orange of Bahia, with Notes on Some Other Little-Known Brazilian Fruits. U.S.D.A. Bull. 445: 1–35 (with P. H. Dorsett and A. D. Shamel).
1917. The Natural Groups of Mangos Cultivated in Florida. Proceedings of the American Pomological Society: 70–80.
1917. The Pollination of the Mango. U.S.D.A. Bull. 542: 1–20.
1919. The Useful Plants of Copán. The American Anthropologist XXI (2): 125–138.
1919. The Avocado in Guatemala. U.S.D.A. Bull. 743: 1–69.
1919. The Avocados of Mexico. A Preliminary Report. Annual Report of the California Avocado Association for 1918–1919: 58–75.
1920. The Colombian Berry, or Giant Blackberry of Colombia. Journal of Heredity XI (5): 195–203.
1920. The Tree Dahlia of Guatemala. Journal of Heredity XI (6): 264–268.
1920. Atlixco. Annual Report of the California Avocado Association for 1919–1920: 24–50.
1920. Manual of Tropical and Subtropical Fruits. The Macmillan Company, New York.
1921. The Pejibaye. Journal of Heredity XII (4): 154–166 (with Otón Jimenez L.).
1921. The Andes Berry. Journal of Heredity XII (9): 386–393.
1921. The Frutilla, or Chilean Strawberry. Journal of Heredity XII (10): 457–465.
1922. The Capulin Cherry. Journal of Heredity XIII (2): 50–62 (with Abelardo Pachano).
1922. The Home of María. Bulletin of the Pan American Union. March: 254–261.
1922. Notes on Peruvian Fruits. Bulletin of the Pan American Union. October: 356–365.
1922. Avocados of the Chota Valley, Ecuador. Annual Report of the California Avocado Association for 1921–1922: 35–39.
1924. Culture de l'Avocatier aux États-Unis. Revue de Botanique Appliquée (Paris). February: 101–106.
1924. The Economic Fruit-Bearing Plants of Ecuador. Contr. from the U.S. National Herbarium 24 (5): 101–134.
1924. Hunting New Fruits in Ecuador. Natural History XXIV (4): 454–466.
1925. Introducción de Plantas Útiles: su importancia para la América Latina. Serie de Agricultura, Unión Panamericana 6: 1–13.

1925. Foreign Service and Agriculture. The American Foreign Service Journal II (11): 365–369; 390.
1926. Round about Bogotá. National Geographic Magazine XLIX (2): 127–160.
1926. El Aguacate: su cultivo y distribución. Serie de Agricultura, Unión Panamericana 16: 329–347.
1926. The Greatest Americans. Unifruitco Magazine. October: 148–157.
1927. Cultivo del Mango en la América Latina. Serie de Agricultura, Unión Panamericana 33: 376–379.
1927. Wild Avocados. Yearbook of the California Avocado Association: 51–54.
1928. The Mangosteen in America. Journal of Heredity XIX (12): 537–545.
1930. Lancetilla Experiment Station. Spanish edition of the Bulletin of the Pan American Union. June: 4–11 (with Alfred F. Butler).
1931. Las Frutas de Guatemala. Anuario del Servicio Técnico de Cooperación Agrícola (Guatemala): 55–93.
1931. The Human Background of Lancetilla. Unifruitco Magazine. August: 19–25 (with Dorothy Popenoe).
1934. Early History of the Avocado. Yearbook of the California Avocado Association: 106–110.
1935. Origin of the Cultivated Races of Avocados. Yearbook of the California Avocado Association: 184–194.
1937. Teocinte in Guatemala. Contributions to American Archaeology, Carnegie Institution of Washington 23: 201–221 (with J. H. Kempton).
1939. Leaf Spot of Bananas. Journal of the Jamaica Agricultural Society. August: 237–241.
1941. Banana Culture around the Caribbean. Tropical Agriculture XVIII (1): 8–12; XVIII (2): 33–38.
1941. The Avocado: A Horticultural Problem. Tropical Agriculture XVIII (1): 3–7.
1941. The Mango: A Study in Systematic Pomology. Tropical Agriculture XVIII (2): 23–25.
1941. Cinchona in Guatemala. Tropical Agriculture XVIII (4): 70–74.
1941. *Cupressus benthamii*: A Neglected Opportunity? Tropical Woods, Yale Univ. School of Forestry 65: 1–4.
1941. Cultivo de la Quina en Guatemala. Boletín de la Dirección General de Agricultura, Guatemala: 1–39.

1942. Grapes for Tropical America. Tropical Agriculture XIX (2): 23–28.
1942. Plant Resources for Honduras. Chronica Botanica VII (5): 217–219.
1943. Cinchona. Boletín Técnico, Departamento Nacional de Agricultura, San José, Costa Rica 42: 1–23.
1946. The Development of Inter-American Cooperation in Agriculture. Bulletin of the Pan American Union LXXX (7): 362–374.
1951. The Story of the Escuela Agrícola Panamericana. The United Fruit Company, Tegucigalpa, Honduras.
1952. Central American Fruit Culture. Ceiba I (5): 269–367.
1956. Tribute to Florida. Ceiba IV (6): 309–315.
1962. Temperate Zone Fruits in the Central American Highlands. Proceedings of the Caribbean Region, American Society for Horticultural Science 6: 3–19 (with Jorge Benítez).
1963. David Fairchild, Plantsman. Bulletin of the Fairchild Tropical Garden 18 (3): 69–80.
1964. Temperate Zone Fruits for Tropical America. Proceedings, Soil and Crop Science Society of Florida 24: 1–12.
1969. The Value of Systematic Pomology in Tropical Fruit Culture. Florida State Horticultural Society 82: 309–313.
1971. Cherimoya, a Bit of Background. California Avocado Society Year Book 54: 133–139.

Bibliography

Adamic, Louis. 1937. The House in Antigua. Harper & Brothers, New York.

Brigham, William T. 1965. Guatemala, the Land of the Quetzal. University of Florida Press, Gainesville, Florida. (Introduction by Wilson Popenoe in a facsimile of the 1887 edition.)

Burbank, Addison. 1939. Guatemala Profile. Coward-McCann, Inc., New York.

Dickson, John D. 1977. Check list and uses of plants in the Wilson Popenoe Botanical Garden, Lancetilla, Honduras. Siatsa. Bull. 6: 1–122.

Doebley, John. 1990. Molecular Evidence and the Evolution of Maize. Economic Botany 44 (3): 6–27.

Dorsett, P. H.; Shamel, A. D.; and Popenoe, Wilson. 1917. The Navel Orange of Bahia with Notes on Some Other Little-known Brazilian Fruits. U.S.D.A. Bull. 445: 1–35.

Dunlap, V. C. 1966. The United Fruit Company and the Lancetilla Experiment Station. Proceedings of the International Symposium on Plant Introduction. Escuela Agrícola Panamericana. Tegucigalpa, Honduras. 33–41.

Escuela Agricola Panamericana. 1946–1990. Annual Reports. Zamorano, Honduras.

Fairchild, David. 1930. Exploring for Plants. The Macmillan Company, New York.

———. 1939. The World Was My Garden. Charles Scribner's Sons, New York.

Fortune Magazine. March 1933. United Fruit. VII (3): 24–38; 120–126.

Frost, Melvin J., and Judd, Ira B. 1970. Plant Explorer of the Americas. Economic Botany 24 (4): 471–478.

Goldblatt, Emily. 1988. Unpublished notes. Wilson Popenoe, Plantsman of the Americas. Arnold Arboretum, Harvard University: 1–23.

Hill, Albert F. 1952. Economic Botany. McGraw-Hill Book Company, Inc., New York.

Kelsey, Vera, and Osborne, Lilly de Jongh. 1939. Four Keys to Guatemala. Funk and Wagnalls Company, New York.

Kempton, J. H., and Popenoe, Wilson. 1937. Teocinte in Guatemala. Contributions to American Archaeology, Carnegie Institution of Washington 23: 201–221.

Mallis, Arnold. 1971. American Entomologists. Rutgers University Press, New Brunswick, New Jersey.

Markham, Albert H. 1917. The Life of Sir Clements R. Markham. John Murray, London.

May, Stacy, and Plaza, Galo. 1958. The United Fruit Company in Latin America. National Planning Association, New York.

McEvoy, J. P. 1947. A Lot of Leaven in a Little Loaf. Pan-American: Magazine of the Americas. July–September: 132–134.

National Research Council. 1989. Lost Crops of the Incas: Little-Known Plants of the Andes with Promise for Worldwide Cultivation. National Academy Press, Washington, D.C.

Popenoe, Dorothy Hughes. 1940. Santiago de los Caballeros de Guatemala. (With an introduction by Thomas Barbour.) Harvard University Press, Cambridge, Massachusetts.

———. 1934. Some Excavations at Playa de los Muertos, Ulua River, Honduras. Maya Research I (2): 61–81.

Popenoe, Paul. 1973. The Date Palm. Field Research Projects, Coconut Grove, Florida.

———. Unpublished, undated manuscript. Popenoe, a Pioneer Family.

Popenoe, Wilson. 1909–1921. Autobiographical Notes. Unpublished manuscript.

———. 1919. The Avocado in Guatemala. U.S.D.A. Bull. 743: 1–69.

———. 1941. Banana Culture around the Caribbean. Tropical Agriculture XVIII (1): 8–12; (2): 33–38.

———. 1946. The Development of Inter-American Cooperation in Agriculture. Bulletin of the Pan American Union LXXX (7): 362–374.

———. 1925. Foreign Service and Agriculture. The American Foreign Service Journal II (11): 365–369; 390.

———. 1921. The Frutilla, or Chilean Strawberry. Journal of Heredity XII (10): 457–465.

———. 1926. The Greatest Americans. Unifruitco Magazine. October: 148–157.

———. 1924. Hunting New Fruits in Ecuador. Natural History XXIV (4): 454–466.

———. 1920. Manual of Tropical and Subtropical Fruits. The Macmillan Company, New York.

———. 1942. Plant Resources of Honduras. Chronica Botanica VII (5): 217–219.

———. 1926. Round about Bogotá. National Geographic Magazine XLIX (2): 127–160.

———. 1951. The Story of the Escuela Agrícola Panamericana. The United Fruit Company, Tegucigalpa, Honduras.

———. 1956. Tribute to Florida. Ceiba IV (6): 309–315.

———. 1919. The Useful Plants of Copán. American Anthropologist 21 (2): 125–138.

———. 1962. Temperate Zone Fruits in the Central American Highlands. Proceedings of the Caribbean Region, American Society for Horticultural Science 6: 3–19 (with Jorge Benítez).

———. 1930. Lancetilla Experiment Station. Spanish edition of the Bulletin of the Pan American Union. June: 4–11 (with Alfred F. Butler).

———. 1931. The Human Background of Lancetilla. Unifruitco Magazine. August: 19–25 (with Dorothy Popenoe).

Rorty, James. 1950. Engineers of World Plenty. Public Affairs Institute, Washington, D.C.

Rosengarten, Frederic, Jr. 1973. The Book of Spices. Pyramid Books, New York.

———. 1944. History of the *Cinchona* Project of Merck & Co., Inc., and Experimental Plantations, Inc. 1934–1943. Merck & Co., Inc., Rahway, New Jersey.

Ryerson, Knowles A. 1977. The World Is My Campus. Regents of the University of California. (Recorded Interview,) Shields Library. The Oral History Center. Davis, California.

Simmonds, N. W. 1966. Bananas. Longman Group Limited, London.

Smiley, Nixon. Unpublished, undated manuscript. Gringo in Guatemala.

———. Miami Herald, column on: June 18, 1950; December 16, 1950; March 18, 1951; March 9, 1952; June 9, 1957; December 8, 1964; March 28, 1965; and April 15, 1965.

Smith, N. J. H.; Williams, J. T.; Plucknett, D. L.; and Talbot, J. In preparation. Tropical Forests and Crop Genetic Resources.

Standley, Paul C. 1929. Lancetilla Experiment Station, Honduras. Unifruitco Magazine. September: 101–104.

Stephens, Clyde S. 1989. Bananeros in Central America. Press Printing Company, Fort Myers, Florida.

The Washington Times. May 24, 1914.

Thompson, J. Eric S. 1958. Thomas Gage's Travels in the New World. University of Oklahoma Press, Norman, Oklahoma.

Uphof, J. C. Th. 1968. The Dictionary of Economic Plants, 2d ed. Verlag von J. Cramer, Würzburg, Germany.

Verdoorn, Frans (ed.). 1945. Plants and Plant Science in Latin America. Chronica Botanica Company, Waltham, Massachusetts.

Washington Evening Star. April 18, 1923.

Wilson, Charles Morrow. 1947. Empire in Green and Gold. Henry Holt and Company.

———. July 1944. School of Pan American Agriculture. Scientific Monthly: 29–36.

Woodger, Elizabeth R. 1983. Wilson Popenoe, American Horticulturist, Educator and Explorer. Huntia 5 (1): 17–23.

INDEX